Numerical Simulations of Turbulent Combustion

Numerical Simulations of Turbulent Combustion

Special Issue Editor

Andrei Lipatnikov

MDPI • Basel • Beijing • Wuhan • Barcelona • Belgrade • Manchester • Tokyo • Cluj • Tianjin

Special Issue Editor
Andrei Lipatnikov
Chalmers University of Technology
Sweden

Editorial Office
MDPI
St. Alban-Anlage 66
4052 Basel, Switzerland

This is a reprint of articles from the Special Issue published online in the open access journal *Fluids* (ISSN 2311-5521) (available at: https://www.mdpi.com/journal/fluids/special_issues/turbulent_combustion).

For citation purposes, cite each article independently as indicated on the article page online and as indicated below:

LastName, A.A.; LastName, B.B.; LastName, C.C. Article Title. *Journal Name* **Year**, *Article Number*, Page Range.

ISBN 978-3-03936-545-6 (Hbk)
ISBN 978-3-03936-546-3 (PDF)

© 2020 by the authors. Articles in this book are Open Access and distributed under the Creative Commons Attribution (CC BY) license, which allows users to download, copy and build upon published articles, as long as the author and publisher are properly credited, which ensures maximum dissemination and a wider impact of our publications.

The book as a whole is distributed by MDPI under the terms and conditions of the Creative Commons license CC BY-NC-ND.

Contents

About the Special Issue Editor vii

Andrei N. Lipatnikov
Numerical Simulations of Turbulent Combustion
Reprinted from: *Fluids* 2020, 5, 22, doi:10.3390/fluids5010022 1

Fernando Luiz Sacomano Filho, Louis Dressler, Arash Hosseinzadeh, Amsini Sadiki, Guenther Carlos Krieger Filho
Investigations of Evaporative Cooling and Turbulence Flame Interaction Modeling in Ethanol Turbulent Spray Combustion Using Tabulated Chemistry
Reprinted from: *Fluids* 2019, 4, 187, doi:10.3390/fluids4040187 5

Ahmed Faraz Khan, Philip John Roberts and Alexey A. Burluka
Modelling of Self-Ignition in Spark-Ignition Engine Using Reduced Chemical Kinetics for Gasoline Surrogates
Reprinted from: *Fluids* 2019, 4, 157, doi:10.3390/fluids4030157 25

Aaron Endres and Thomas Sattelmayer
Numerical Investigation of Pressure Influence on the Confined Turbulent Boundary Layer Flashback Process
Reprinted from: *Fluids* 2019, 4, 146, doi:10.3390/fluids4030146 37

Andrei N. Lipatnikov, Shinnosuke Nishiki and Tatsuya Hasegawa
Closure Relations for Fluxes of Flame Surface Density and Scalar Dissipation Rate in Turbulent Premixed Flames
Reprinted from: *Fluids* 2019, 4, 43, doi:10.3390/fluids4010043 57

Rixin Yu and Andrei N. Lipatnikov
DNS Study of the Bending Effect Due to Smoothing Mechanism
Reprinted from: *Fluids* 2019, 4, 31, doi:10.3390/fluids4010031 69

Ahmad Alqallaf, Markus Klein and Nilanjan Chakraborty
Effects of Lewis Number on the Evolution of Curvature in Spherically Expanding Turbulent Premixed Flames
Reprinted from: *Fluids* 2019, 4, 12, doi:10.3390/fluids4010012 83

Arne Heinrich, Guido Kuenne, Sebastian Ganter, Christian Hasse and Johannes Janicka
Investigation of the Turbulent Near Wall Flame Behavior for a Sidewall Quenching Burner by Means of a Large Eddy Simulation and Tabulated Chemistry
Reprinted from: *Fluids* 2018, 3, 65, doi:10.3390/fluids3030065 107

About the Special Issue Editor

Andrei Lipatnikov received his Ph.D. in Molecular and Chemical Physics from Moscow Institute of Physics and Technology in 1987. Subsequently, he was employed by that Institute until he was invited to join the Department of Thermo and Fluid Dynamics at Chalmers University of Technology as a guest scientist in 1996. In May 1998, he was permanently employed as a researcher at the same department. In August 2000, the School of Mechanical and Vehicular Engineering accepted Dr. Lipatnikov as a docent. In July 2017, he was appointed a research professor. His academic activities concern the modeling of burning of gaseous mixtures in turbulent and laminar flows, pollutant formation in flames, autoignition of premixed reactants, thermo-acoustic instabilities, fuel sprays, and numerical simulations of turbulent flames in laboratory burners and internal combustion engines. He has authored a monograph, four invited book chapters, and about 280 scientific contributions, including 107 original journal papers and five review articles published by *Progress in Energy and Combustion Science and Annual Review of Fluid Mechanics*.

Editorial

Numerical Simulations of Turbulent Combustion

Andrei N. Lipatnikov

Department of Mechanics and Maritime Sciences, Chalmers University of Technology, 412 96 Gothenburg, Sweden; andrei.lipatnikov@chalmerse.se

Received: 1 February 2020; Accepted: 5 February 2020; Published: 10 February 2020

Turbulent burning of gaseous fuels is widely used for energy conversion in stationary power generation, e.g., gas turbines, land transportation, e.g., piston engines, and aviation, e.g., aero-engine afterburners. Nevertheless, fundamental understanding of turbulent combustion is still limited, because it is a highly non-linear and multiscale process that involves various local phenomena and thousands (e.g., for gasoline-air mixtures) of chemical reactions between hundreds of species, including a number of reactions that control emissions from flames. Therefore, there is a strong need for elaborating high fidelity, advanced numerical models and methods that (i) will catch complex combustion chemistry and the governing physical mechanisms of flame-turbulence interaction and, consequently, (ii) will make turbulent combustion computations an efficient predictive tool for applied research. In particular, such computations are required to facilitate development of a new generation of ultra clean and highly efficient internal combustion engines that will allow the society to properly respond to current environmental and efficiency challenges. The goal of this special issue is to provide a forum for recent developments in such numerical models and methods. The special issue contains papers aimed at (i) developing and validating high fidelity models and efficient numerical methods for Computational Fluid Dynamics research into turbulent, complex-chemistry combustion in laboratory burners and in engines or (ii) improving fundamental understanding of flame-turbulence interaction by analyzing data obtained in unsteady multi-dimensional numerical simulations.

Khan et al. [1] report results of a joint experimental and numerical study of chemical processes that cause autoignition of a fuel-air mixture and, in particular, knock in Spark Ignition engines. More specifically, ignition delay times computed invoking three reduced (semi-detailed) chemical mechanisms for different gasoline surrogates are compared with timing of knock onset measured for a wide range of temperatures and pressures. Obtained results indicate that the studied chemical mechanisms and surrogate properties can feasibly be used in the calculation of gasoline autoignition in a Spark Ignition engine, with the computed ignition delay time being sensitive to the choice of a mechanism and/or surrogate.

Endres and Sattelmayer [2] present results of large eddy simulations performed by allowing for complex combustion chemistry at various pressures. The goal of the study is to numerically explore boundary layer flashback in a confined combustion chamber when burning hydrogen-air mixtures. Results show that while the turbulent flame speed at conditions close to flashback decreases with increasing pressure, the flashback propensity is increased by the pressure. This finding indicates that a single quantity such as the turbulent flame speed is a poor indicator for the onset of boundary layer flashback. The flashback is a complex process affected by the flame speed, the flame thickness, the quenching distance, and the local separation zone size. Moreover, the computed results show that the underlying assumptions of the boundary layer theory are not satisfied under conditions of the study. For instance, application of one-dimensional pressure approximations results in overestimating the pressure increase ahead of the flame.

To perform large eddy simulations of turbulent burning of ethanol sprays, Filho et al. [3] develop a modeling strategy that allows for complex combustion chemistry by combining Flamelet Generated Manifolds (GFM) and Artificially Thickened Flame (ATF) approach extended by the authors to take into account enthalpy variations due to evaporative cooling effects. Ethanol droplets are tracked using an Euler–Lagrangian approach and applying an evaporation model to allow for the inter-phase non-equilibrium. Numerical results are validated using experimental data obtained from flame EtF5 of the Sydney diluted spray flame burner. Moreover, a parametric numerical study is performed to assess magnitudes of effects due to evaporation cooling and wrinkling of flame surface by turbulent eddies, with the latter effect being of more importance.

Heinrich et al. [4] apply large eddy simulation to study another problem, i.e., flame-wall interaction, which can promote pollutant formations and increase heat losses, thus, lowing efficiency of an internal combustion engine. Similar to Ref. [3], complex combustion chemistry and flame-turbulence interaction are taken into account adopting the FGM and ATF approaches, respectively. The numerical model is validated using experimental data on sidewall quenching of turbulent flames, obtained recently in Darmstadt. The validation study shows that the adapted numerical approach can handle sidewall quenching of turbulent flames. Moreover, in the paper, the computed instantaneous 3D fields are analyzed and three different scenarios are revealed. These are: an upstream, a downstream and a jump-like upstream movement of the flame. In the third case, the flame behaves locally like a head-on quenching flame and the highest heat fluxes are calculated.

Alqallaf et al. [5] analyze direct numerical simulation data obtained from expanding, statistically spherical turbulent premixed flames characterized by three different Lewis numbers, i.e., Le = 0.8, 1.0, and 1.2, with all other things being equal. By processing the data, various terms in the transport equation for the local curvature of the instantaneous flame surface are evaluated and terms due to curl of vorticity and normal strain rate gradients are found to play the most important roles in the studied transport equation in all three cases. In the case of Le = 0.8, the net contribution of the considered terms acts to augment turbulence-induced wrinkles on the flame surface. In two other cases of Le = 1.0 and 1.2, flame propagation tends to smoothen the flame surface. These findings shed a new light on the influence of the Lewis number on turbulent burning rate.

Yu and Lipatnikov [6] compare direct numerical simulation data computed by studying two model problems relevant to premixed turbulent combustion. These are (i) motion of a self-propagating interface in a constant-density turbulence and (ii) propagation of a reaction wave of a finite thickness in a constant-density turbulence. Both data sets are obtained from statistically the same turbulence. In the former case, the computed mean speed of the interface is proportional to the rms turbulent velocity u', whereas the dependence of the mean wave speed on u' shows bending, which is more pronounced in the case of a higher diffusivity of the reactant, i.e., a larger local wave thickness. Analysis of the data indicates that the bending effect is controlled by a decrease in the rate of an increase in the reaction-zone-surface area with increasing u'. This decrease stems from inefficiency of small-scale turbulent eddies in wrinkling the reaction-zone surface, because such small-scale wrinkles characterized by a high local curvature are efficiently smoothed out by molecular diffusion within the reaction wave.

Lipatnikov et al. [7] suggest new closure relations for turbulent scalar fluxes of flame surface density and scalar dissipation rate in the corresponding transport equations. These closure relations are validated by analyzing direct numerical simulation data obtained from three statistically stationary, one-dimensional, planar, weakly turbulent premixed flames characterized by three different density ratios. The models predict the fluxes reasonably well without using any tuning parameter and can yield both gradient and countergradient fluxes in different zones of the mean flame brushes, with the zone sizes depending on the density ratio.

I thank all of the authors for submitting their manuscripts for this special issue. I also thank all of the reviewers for their time and valuable comments increasing the quality of the published papers.

Conflicts of Interest: The author declares no conflict of interest.

References

1. Khan, A.F.; Roberts, P.J.; Burluka, A.A. Modelling of self-ignition in Spark-Ignition engine using reduced chemical kinetics for gasoline surrogates. *Fluids* **2019**, *4*, 157. [CrossRef]
2. Endres, A.; Sattelmayer, T. Numerical investigation of pressure influence on the confined turbulent boundary layer flashback process. *Fluids* **2019**, *4*, 146. [CrossRef]
3. Filho, F.L.S.; Dressler, L.; Hosseinzadeh, A.; Sadiki, A.; Filho, G.C.K. Investigations of evaporative cooling and turbulence flame interaction modeling in ethanol turbulent spray combustion using tabulated chemistry. *Fluids* **2019**, *4*, 187. [CrossRef]
4. Heinrich, A.; Kuenne, G.; Ganter, S.; Hasse, C.; Janicka, J. Investigation of the turbulent near wall flame behavior for a sidewall quenching burner by means of a large eddy simulation and tabulated chemistry. *Fluids* **2018**, *3*, 65. [CrossRef]
5. Alqallaf, A.; Klein, M.; Chakraborty, N. Effects of Lewis number on the evolution of curvature in spherically expanding turbulent premixed flames. *Fluids* **2019**, *4*, 12. [CrossRef]
6. Yu, R.; Lipatnikov, A.N. DNS Study of the bending effect due to smoothing mechanism. *Fluids* **2019**, *4*, 31. [CrossRef]
7. Lipatnikov, A.N.; Nishiki, S.; Hasegawa, T. Closure relations for fluxes of flame surface density and scalar dissipation rate in turbulent premixed flames. *Fluids* **2019**, *4*, 43. [CrossRef]

© 2020 by the author. Licensee MDPI, Basel, Switzerland. This article is an open access article distributed under the terms and conditions of the Creative Commons Attribution (CC BY) license (http://creativecommons.org/licenses/by/4.0/).

Article

Investigations of Evaporative Cooling and Turbulence Flame Interaction Modeling in Ethanol Turbulent Spray Combustion Using Tabulated Chemistry

Fernando Luiz Sacomano Filho [1,*], Louis Dressler [2], Arash Hosseinzadeh [2], Amsini Sadiki [2] and Guenther Carlos Krieger Filho [1]

1 Laboratory of Environmental and Thermal Engineering, University of São Paulo, São Paulo 05508-030, Brazil; guenther@usp.br
2 Institute for Energy and Power Plant Technology, Technische Universität Darmstadt, 64287 Darmstadt, Germany; dressler@ekt.tu-darmstadt.de (L.D.); zadeh@ekt.tu-darmstadt.de (A.H.); sadiki@ekt.tu-darmstadt.de (A.S.)
* Correspondence: sacomano@ekt.tu-darmstadt.de

Received: 30 June 2019; Accepted: 29 October 2019; Published: 31 October 2019

Abstract: Evaporative cooling effects and turbulence flame interaction are analyzed in the large eddy simulation (LES) context for an ethanol turbulent spray flame. Investigations are conducted with the artificially thickened flame (ATF) approach coupled with an extension of the mixture adaptive thickening procedure to account for variations of enthalpy. Droplets are tracked in a Euler–Lagrangian framework, in which an evaporation model accounting for the inter-phase non-equilibrium is applied. The chemistry is tabulated following the flamelet generated manifold (FGM) method. Enthalpy variations are incorporated in the resulting FGM database in a universal fashion, which is not limited to the heat losses caused by evaporative cooling effects. The relevance of the evaporative cooling is evaluated with a typically applied setting for a flame surface wrinkling model. Using one of the resulting cases from the evaporative cooling analysis as a reference, the importance of the flame wrinkling modeling is studied. Besides its novelty, the completeness of the proposed modeling strategy allows a significant contribution to the understanding of the most relevant phenomena for the turbulent spray combustion modeling.

Keywords: spray combustion; evaporative cooling; flame surface wrinkling modeling; thickened flame; flamelet generated manifold

---

1. Introduction

Turbulent spray combustion features the unsteady interaction of many strongly coupled physical phenomena varying in a broad range of time and length scales. The complexity resulting from such interacting phenomena can be appraised with the studies presented for instance by Jenny et al. [1], Gutheil [2], and Sacomano Filho [3]. Heat and mass transfer between phases and the effects of the turbulence on a flame, as well as their reciprocal, stand out among these coupled phenomena. Although both interactions are strongly coupled, they are not always comprehensively addressed in numerical simulations. The cooling down of the gas mixture caused by the evaporation process, namely the evaporative cooling (EC), is often neglected [4–8] or simplified [9,10] in the context of tabulated chemistry. Herein, the consideration of heat losses is not trivial and typically involves high computational efforts [11–13]. Commonly, the low volume fraction of diluted sprays are used to justify simulations without heat losses [14,15]. Nevertheless, as analyzed in [16] for one-dimensional flames propagating in droplet mists, the inclusion of the evaporative cooling is quite relevant in the spray combustion modeling. Yet the importance of the proper modeling of the turbulence-flame interaction is a well known topic in the spray combustion community [1,4,6–10,13,17–21]. In view of the application of the artificially thickened flame (ATF) approach, many works have been using a similar value for the exponent of the power-law function used to model the flame surface wrinkling (FSW), namely 0.5. Recent investigations presented in [3,21] show that the characterization of the turbulence-spray flame interaction can be significantly improved according to the value choice of such a model parameter. However, this has not been consistently analyzed with spray flames accounting for the evaporative cooling up to the present study.

An initial approach to account for evaporative cooling effects is to consider heat losses in the mixture without including modifications in the chemistry table [9,10]. According to this formalism, influences of heat losses concentrate on the computation of droplet evaporation rates which indirectly interfere with the combustion process. As evaporation rates change, the mixture composition arriving at the reaction zone also changes. However, the influences of heat losses are not strictly accounted for in the chemistry. Knudsen et al. [13] proposed a low computational cost strategy to include heat losses exclusively caused by droplets in the flamelet-progress variable approach. They compute freely propagating and counter-flow flamelets to generate two tables (one based on freely propagating and another based on counter-flow flamelets) which would be chosen during the simulation evolution by means of a flame index. Instead of considering flamelets in diverse enthalpy levels, this variable is constrained with the mixture fraction by assuming that all the fuel existing in the fresh mixture comes from liquid evaporation. As a result, no enthalpy must be transported. This simplification can be applied in cases where the unitary Lewis number (Le) approach is adopted. The validity of this method for the description of the reaction evolution in mixtures richer than the mixture fraction corresponding to saturated vapor is not comprehensively investigated. As mentioned by van Oijen and de Goey [11], freely propagating flamelets become unphysical below a certain level of enthalpy. Therefore, care must be taken in order to suitably describe rich reactions with this method. Recently, Olguin and Gutheil [22] presented a new procedure to tabulate the chemistry involving spray flames. It does not only account for the effects of heat losses, but also to the presence of droplets in the reaction zone. Hu et al. [23] show that this method is quite promising to address turbulent spray flames. However, some obstacles can also be encountered there. For instance, the increase of table dimensions that request more storage and computational efforts, as well as the difficulty to find correspondence of the droplet trajectories in the counter-flow setup used to construct such a table [16] with the other found in a multi-dimensional flame. Similar to [22], Luo et al. [24] and Franzelli et al. [25] propose a method based on spray flamelets to characterize spray flames in view of tabulated chemistry. Despite being considered, effects of the evaporative cooling in [24] are limited since droplets are injected at constant (boiling point) temperature. As in [22], such limitation is not found in [25]. However, this last study considers a monodisperse droplet cloud, which is a more restrictive condition for the dispersed phase when compared with [22,24].

The ATF model stands out among methods based on deterministic approaches to characterize the turbulence-flame interaction in turbulent spray flames [9,10,17,20,21]. This model showed a great capability to predict the behavior of turbulent spray flames in the Eulerian–Eulerian two-phase flow approach (e.g., [9,17]) as well as in the Euler–Lagrangian framework (e.g., [10,20,21]). Initially proposed to model laminar premixed flames [26], the ATF was gradually expanded to address turbulent premixed [27,28] and stratified flames [29]. In view of turbulent spray combustion, a further extension to include effects of the dispersed phase in the Euler–Lagrangian context are recently presented in [21]. A common characteristic to all of these preceding works involving turbulent flames (i.e., [9,10,20,21]) refers to the usage of a global and presumed definition of the exponent of the power-law function (β) proposed by Charlette et al. [28] to model the FSW. According to the analysis conducted for premixed flames in [30], no universal value exists for this parameter, which may be determined by adjusting with experimental data or by some modeling approach (e.g., the dynamic method proposed by Charlette et al. [30]). Preliminary studies accounting for the dynamic modeling of such a model parameter (see [3]) showed promising results for the characterization of turbulence-spray flame interactions.

However, further investigations are still necessary to turn this approach more consistent for general turbulent spray flame simulations. On the other hand, analysis of the influence of the model parameter β are needed for spray flames, which have not been comprehensively addressed in cases where heat losses are considered.

In the present work, the effectiveness of the evaporative cooling and the FSW modeling to characterize a diluted turbulent spray flame is analyzed. The chemistry tabulation method proposed by van Oijen and de Goey [11] and comprehensively tested for spray flames in [16] is employed here. Namely, effects of the evaporative cooling interfere with reaction rates, and the consideration of heat losses are not limited to the evaporative cooling in the flamelet generated manifold (FGM) table. Simulations are conducted following the large eddy simulation (LES) approach coupled with the ATF method. Besides the novelty associated with such a consistent consideration of heat losses in a turbulent spray flame, the mixture adaptive thickening approach [21] is extended to include effects of enthalpy variations. The flame EtF5 of the Sydney diluted spray flame burner [31] is selected as a benchmark to the subsequent analysis. Detailed chemistry effects are included by 57 species and 379 intermediary reactions mechanism of ethanol-air proposed by Marinov [32] by means of the FGM methodology. The unsteadiness arising from the turbulent dispersion of evaporating droplets are captured by a Euler–Lagrangian spray module relying on the LES approach. An evaporation model accounting for the inter-phase non-equilibrium is applied to describe the droplet evaporation process. A parametric study has been performed to evaluate the effectiveness of the evaporative cooling and the turbulence flame interaction modeling consideration.

The remaining structure of this paper is divided into three parts. An overview of the theoretical background is described in Section 2. In Section 3, simulation results are presented. Analyses of effects of the evaporative cooling and its coupling with the flame surface wrinkling modeling are systematically addressed. In the last part, final remarks and the main conclusions are summarized.

2. Materials and Methods

A Euler–Lagrangian approach was adopted to represent the two-phase flow. Herein, a full inter-phase two-way coupling was accounted for. Carrier gas-phase quantities were interpolated into droplets positions, while influences of the dispersed phase were introduced through source terms in computational cells.

2.1. Gas Phase

The turbulent motions of the carrier phase are described in the LES context following a variable-density low Mach number formulation. According to this approach, mass and momentum equations are described by

$$\frac{\partial \overline{\rho}}{\partial t} + \frac{\partial \overline{\rho} \widetilde{u}_j}{\partial x_j} = \overline{S_m} \quad (1)$$

$$\frac{\partial \overline{\rho} \widetilde{u}_i}{\partial t} + \frac{\partial \overline{\rho} \widetilde{u}_i \widetilde{u}_j}{\partial x_j} = \frac{\partial}{\partial x_j}\left(2\overline{\mu}\widetilde{S}_{ij} - \frac{2}{3}\overline{\mu}\frac{\partial \widetilde{u}_k}{\partial x_k}\delta_{ij} - \overline{\rho}\tau^{sgs}_{ij}\right) - \frac{\partial \overline{p}}{\partial x_i} + \overline{\rho}g_i + \overline{S_{u,i}}. \quad (2)$$

The dependent filtered variables were obtained from spatial filtering as $\psi = \widetilde{\psi} + \psi''$ with $\widetilde{\psi} = \overline{\rho\psi}/\overline{\rho}$. Over-bars and tildes express spatially filtered and density-weighted filtered values with a filter width Δ_{mesh}, respectively, while double prime represents sub-grid scale (SGS) fluctuations. ρ is the mixture density, t time, u_j components of velocity in j ($j = 1, 2, 3$) direction, p pressure, x_j Cartesian coordinate in j direction, μ the dynamic viscosity, g_i the component of gravitational acceleration, δ_{ij} the Kronecker's delta, and S_{ij} the strain rate. The term $\overline{S_m}$ corresponds to the introduction of mass from the droplets into the carrier phase, while $\overline{S_{u,i}}$ is the source term of momentum due to the presence of the dispersed phase. Both follow the implementations presented by Chrigui et al. [33]. The SGS stress tensor τ^{sgs}_{ij} is closed by means of the Smagorinsky model with the dynamic procedure of Germano et al.

[34]. More details about the mathematical treatment given here to mass and momentum equations can be found in Sacomano Filho et al. [21].

2.1.1. Mixture Formation and Combustion Modeling

In order to account for the evaporative cooling and general heat exchanges, three scalar quantities are used to characterize the mixture following the FGM method: the mixture fraction Z, the reaction progress variable Y_{pv}, and the absolute enthalpy of the gas mixture h. The transport equation for them can be written in terms of a general variable ψ within the ATF modeling as

$$\frac{\partial(\bar{\rho}\tilde{\psi})}{\partial t} + \frac{\partial(\bar{\rho}\tilde{\psi}\tilde{u}_j)}{\partial x_j} = \frac{\partial}{\partial x_j}\left[\left(FE^*_\Delta \frac{\bar{\mu}}{\sigma_\psi} + (1-\Omega)\frac{\mu_t}{\sigma_{t,\psi}}\right)\frac{\partial\tilde{\psi}}{\partial x_j}\right] + \frac{E^*_\Delta}{F}\overline{\dot{\omega}_\psi} + \overline{S_\psi}, \tag{3}$$

where μ_t is the turbulent viscosity. For the transport equations of Z and Y_{pv}, σ_ψ and $\sigma_{t,\psi}$ respectively represent the laminar Sc and the turbulent Sc_t Schmidt numbers ($Sc = Sc_t = 0.7$), while for h both respectively correspond to the laminar Pr and the turbulent Pr_t Prandtl numbers ($Pr = Pr_t = 0.7$). It is important to highlight that, with these values for Sc and Pr the unitary $Le = Pr/Sc$ approach is maintained. The quantity F corresponds to the thickening factor, E^*_Δ to the efficiency function, and Ω denotes the flame sensor. Details about these quantities are addressed below.

The term $\dot{\omega}_\psi = \dot{\omega}_\psi\left(\widetilde{Z}, \widetilde{Y_{pv}}, \widetilde{h}\right)$ corresponds to the reaction rate for the Y_{pv} whereas it is set to zero for the mixture fraction and absolute enthalpy equation. Similarly to the reaction progress variable source term, $\bar{\rho} = \rho\left(\widetilde{Z}, \widetilde{Y_{pv}}, \widetilde{h}\right)$ and $\bar{\mu} = \mu\left(\widetilde{Z}, \widetilde{Y_{pv}}, \widetilde{h}\right)$ are obtained from the employed FGM table. Therefore, these are expressed as functions of the transported scalar quantities Z, Y_{pv}, and h. The source term $\overline{S_\psi}$ consists of the source of vapor introduced by the dispersed phase in the transport equation for Z, specifically $\overline{S_Z} = \overline{S_m}$. Considering that the mass fraction of fuel is not present in the combination used to define the reaction progress variable (see Equation (15)), as well as no isolated droplet burning model is included in the employed approach, $\overline{S_\psi}$ does not contribute (therefore it is set to zero) for the transport of Y_{pv}. For the transport equation of h, $\overline{S_h}$ is given by

$$\overline{S_h} = \sum_{p=1}^{N} \frac{N_p}{V}\left[m_{d,p}\left(\int_{T_{ref}}^{T_d^t} c_l dT - \int_{T_{ref}}^{T_d^{t+\Delta t}} c_l dT\right) + \dot{m}_{d,p}\left(h_f - L_v\right)\right], \tag{4}$$

where $\dot{m}_{d,p}$ denotes the mass of vapor released by the parcel p into the control volume V, $m_{d,p}$ is the droplet mass, N_p the number of real droplets in parcel p, N the total number of tracked parcels, c_l the specific sensible heat of liquid, T_{ref} a reference temperature (298 K), T_d^τ the droplet temperature at time step τ, h_f the formations enthalpy, and L_v the heat of vaporization. More description about how the phase coupling source terms are computed in the ATF context can be found in [3,21].

Following the ATF method, the flame thickening is performed by means of a dynamic procedure. Accordingly, only the flame region is thickened and no interferences of the ATF with the pre-vaporization zone occur. The flame sensor Ω used by Aschmoneit [35] to simulate partially premixed flames is employed here. The quantity $\dot{\omega}_{pv,max}\left(\widetilde{Z}, \widetilde{h}\right)$ in Equation (5) is associated to the maximum value of the source term $\dot{\omega}_{pv}\left(\widetilde{Z}, \widetilde{Y_{pv}}, \widetilde{h}\right)$ at the same mixture composition and enthalpy level.

$$\Omega = \min\left(1.0, \left(0.25\Omega_c + 0.75 \times 2.0\frac{\dot{\omega}_{pv}\left(\widetilde{Z}, \widetilde{Y_{pv}}, \widetilde{h}\right)}{\dot{\omega}_{pv,max}\left(\widetilde{Z}, \widetilde{h}\right)}\right)\right), \tag{5}$$

where Ω_c is the flame sensor proposed by Durand and Polifke [36] as implemented in [29]. In this context, a progress variable in its normalized form is needed, which is defined as $c = Y_{pv}/Y_{pv}^{eq}$. Herein, Y_{pv}^{eq} denotes the equilibrium value of Y_{pv} for a specific mixture composition.

In order to facilitate the access of $\dot{\omega}_{pv,max}$ during the simulation time, an extra look-up table is stored with its values mapped on $\widetilde{Z}$ and $\widetilde{h}$. According to the dynamic method, the thickening factor F is defined as

$$F = 1 + (F_{max} - 1)\Omega. \tag{6}$$

in which F_{max} is defined according to an extension of the mixture adaptive thickening proposed in [21] to account for enthalpy variations as

$$F_{max} = \max\left(1, \frac{\Delta_{mesh}}{\Delta_{x,max}}\right) = \max\left(1, \frac{V^{\frac{1}{3}}}{\Delta_{x,max}}\right) \tag{7}$$

in which V denotes the cell volume and $\Delta_{x,max}$ is the maximum cell size necessary to capture the laminar flame speed with less than 15 % error in one-dimensional simulations [29]. In agreement with Kuenne et al. [29], $\Delta_{x,max} \approx 0.3\delta_l^0(\widetilde{Z}, \widetilde{h})$. Additionally, two safety factors are also considered to this maximum cell size definition. The first of $1/\sqrt{3}$ is included to account for the worst scenario of flame propagating in cell diagonal direction for three-dimensional cases, the second $1/\alpha$ is used to account for the non-uniformity of cells in F. Applying both safety factors, the maximum cell size is defined here as

$$\Delta_{x,max} = \frac{0.3}{\alpha\sqrt{3}}\delta_l^0(\widetilde{Z}, \widetilde{h}), \tag{8}$$

with $\alpha = 1.0$, since cubic cells are employed in the jet core region (see Section 2.3).

To recover the modifications introduced by the ATF on the turbulence-flame interaction while including the effects of the unresolved FSW, the efficiency function E_Δ follows the fractal model. It is written as

$$E_\Delta = \left(\frac{\Delta}{\delta_l^0}\right)^\beta, \tag{9}$$

where $\Delta = F \cdot \delta_l$ is the combustion LES filter width, δ_l^0 is the laminar flame thickness defined from the temperature gradient, β a model exponent, and δ_l an equivalent laminar flame thickness in terms of Gaussian LES filter as proposed in [37] (see more details in the sequel). This is a reasonable approach in the simulations presented here, since we could notice that

$$\Gamma\frac{u'_\Delta}{s_l} > \max\left(\frac{\Delta}{\delta_l^0} - 1, 0\right), \tag{10}$$

already for flames presenting a smaller Reynolds number (Re) in [3]. Considering the minimum operator in

$$E_\Delta = \left(1 + \min\left[\max\left(\frac{\Delta}{\delta_l^0} - 1, 0\right), \Gamma\frac{u'_\Delta}{s_l}\right]\right)^\beta, \tag{11}$$

this inequality shows that the saturation holds.

Equation (9) is connected with the efficiency function E_Δ^* by means of

$$E_\Delta^* = 1 + (E_\Delta - 1)\Omega. \tag{12}$$

Equation (12) is used to enforce that $E_\Delta^* \to 1$ outside the flame region, since an efficiency function is no longer necessary in this part of the domain. Hence, the term $FE_\Delta^*\left(\overline{\mu}/\sigma_\psi\right)$ in Equation (3) reduces to $\overline{\mu}/\sigma_\psi$, whereas $(1 - \Omega)\mu_t/\sigma_{t,\psi}$ becomes $\mu_t/\sigma_{t,\psi}$. Since $\sigma = \sigma_t = 0.7$, both terms are merged in which the molecular viscosity is added with the turbulent one and the effective viscosity $\mu_{eff} = \overline{\mu} + \mu_t$ is obtained. In contrast to this scenario, within the flame region $\Omega \to 1$ and the term $(1 - \Omega)\mu_t/\sigma_{t,\psi}$ is canceled, where molecular and turbulent transport in the flame are modeled by $FE_\Delta\left(\overline{\mu}/\sigma_\psi\right)$.

The determination of Δ follows a formulation that can be directly applied for future implementations accounting for the dynamic modeling of β. Accordingly, the equivalent laminar flame thickness used to define the LES combustion filter is given by $\delta_l = \zeta \delta_l^0$ (in agreement with [37]), in which ζ is an adjusting factor to enforce that the efficiency function goes to one for laminar flames. Values of ζ are supposed to vary according to the mixture state, however only the variations associated with the composition are accounted for in this work. As a result, the actual formulation of the mixture adaptive thickening which accounts for enthalpy variations is considered to achieve

$$\Delta = \delta_l F_{max} = \zeta \delta_l^0 F_{max} = \zeta \delta_l^0(\widetilde{Z}, \widetilde{h}) \left[\max\left(1.0, \alpha \frac{\sqrt{3}}{0.3} \frac{V^{\frac{1}{3}}}{\delta_l^0(\widetilde{Z}, \widetilde{h})}\right) \right]. \tag{13}$$

In regions where the cell sizes deliver F greater than 1, Equation (13) reduces to

$$\Delta = \zeta \left(\alpha \frac{\sqrt{3}}{0.3} V^{\frac{1}{3}} \right). \tag{14}$$

2.1.2. Chemistry

To construct our non-adiabatic FGM table, the strategy proposed by van Oijen and de Goey [11] is applied. Particularly, the reference table described in [16] is employed. It is based on the combination of freely propagating adiabatic flamelets for a mixture composition span of $Z \in [0.050, 0.194]$ with fresh gas temperature varying from 250 K to 900 K with burner stabilized and extrapolated flamelets. The maximum value of this temperature determines the upper enthalpy value, while its minimum defines the lowest temperature (T_{low}) allowed in the computational domain. Interpolations are done [38] between the limiting values of Z up to pure air and ethanol at 300 K, which corresponds to the ambient temperature of the investigations performed in this work. To compute both kinds of flamelets, the chemical mechanism proposed by Marinov [32] is employed. It represents the oxidation of ethanol in the air by means of 57 species and 379 intermediate reactions.

The inclusion of burner-stabilized and extrapolated flamelets in our table allows a more universal description of the heat losses. Burner-stabilized flamelets mimic premixed flames burning attached to a porous medium with controlled temperature. By keeping the temperature of this medium constant and adjusting the inflow velocity, heat losses can be introduced in the domain resulting in different enthalpy levels for the flame. In this particular step, the temperature is set to the same value as for the unburned gas of the coldest adiabatic flamelet. When the lowest enthalpy levels with the burner stabilized flamelets is reached, the burnt gas is still far away from T_{low}. To cover the remaining range of physical states, step-wise extrapolations of the thermo-chemical data are done to the lowest enthalpy level. This is computed by assuming the equilibrium mixture composition at T_{low}. Finally, once a manifold is generated, the look-up table is build up over the controlling variables Z, Y_{pv}, and h. Following the strategy of Ketelheun et al. [12], an additional table is created to facilitate the establishment of the boundary conditions in terms of temperature.

The definition of Y_{pv} follows the combination of mass fractions presented in [16]. This was empirically defined in view of a valid progress variable for the full range of enthalpy levels encountered in the employed table. Namely, a monotonically evolving variable from fresh to burnt gases along all flamelets used to construct the employed FGM table. The Y_{pv} can be written as

$$Y_{pv} = \frac{1}{M_{CO_2}} Y_{CO_2} + \frac{1}{2.5 M_{H_2O}} Y_{H_2O} + \frac{1}{1.5 M_{CO}} Y_{CO}, \tag{15}$$

where M_k is the molar mass of species k. For more details about the definition of a progress variable in FGM context, the reader is referred to [11]).

2.2. Liquid Phase

The computation of the motion of non-rotating droplets considers only drag and buoyancy forces. Once that the density ratio of liquid ethanol and the gaseous mixture has an order of 10^3, complementary forces are neglected.

Heat and mass exchanges are computed following the model proposed by Miller et al. [39] with the correction procedure derived in [21] to include effects of the flame thickening onto the dispersed phase. Both phenomena are described by

$$\frac{dT_p}{dt} = \frac{1}{F}\left[\frac{f_2 Nu}{3Pr}\left(\frac{\theta_1}{\tau_p}\right)(T - T_p) + \left(\frac{L_V}{c_l}\right)\frac{\dot{m}_p}{m_p}\right], \qquad (16)$$

and

$$\frac{dm_p}{dt} = -\frac{1}{F}\frac{Sh}{3Sc}\left(\frac{m_p}{\tau_p}\right)H_M, \qquad (17)$$

with T the temperature, Nu the Nusselt number, f_2 a correction factor due to evaporation computed as in [39], Pr the Prandtl number, θ_1 the ratio of the gaseous and liquid specific sensible heat (c_p/c_l), $\tau_p = \rho_p d_p^2/18\mu$ expresses the particle relaxation time, $\dot{m}_p = dm_p/dt$, Sh the Sherwood number, and H_M represents the specific driving potential for mass transfer according to [39]. It is important to highlight that vapor pressures, necessary to determine H_M are computed here with the Wagner equation as in [40].

For the vast majority of droplets tracked in our simulations, the relaxation time exceeded the scales within the subgrid. As a consequence, SGS turbulent dispersion and micro-mixing modeling for the evaporation process had little effect.

2.3. Experimental Configuration and Numerical Setup

The flame EtF5 of the Sydney diluted spray flame burner [31] is used as a benchmark for the performed LES. This flame belongs to a series of flames (EtF1, EtF2, EtF5, and EtF7), where the premixed flame mode is dominant. This is an important feature in view of the chosen modeling approach, which has been originally developed for premixed flames. EtF5 was produced by a round jet burner with a diameter D of 10.5 mm fueled with ethanol. The flame stabilization was achieved with a series of pilot flames disposed of in a concentric arrange to the main jet nozzle with an annular diameter of 25 mm. A co-flow surrounded both streams with an annular diameter of 104 mm and a velocity of 4.5 m/s, which corresponds to the air velocity of the wind tunnel where the burner is mounted. The bulk velocities of main jet and pilot were 48 m/s and 11.6 m/s, respectively. At the nozzle exit, a mixture of droplets, air, and fuel vapor was formed with a Re of 38,200 and an equivalence ratio $\phi = 0.15$ for the gaseous phase, which is found below the lower flammability limit (ϕ_{low}).

The computational domain is constructed with a Cartesian grid in order to keep the cell growth rate smooth and to avoid large values of the factors F and α (see Equations (3) and (14), respectively). Specifically, mean values of F are about 7.0 but no bigger than approximately 9.0, while $\alpha = 1.0$ due to the cubic cell arrangement in the jet core region. When compared to the "O-grid" arrangement (as employed in [3]), which would be an alternative to run in the used block-structured code, the Cartesian arrangement avoids the high cell diagonal growth rate. Figure 1 presents a sketch of the employed grid, which is composed of approximately 5.4 million cells distributed over 468 blocks. In this figure black lines represent block limits, whereas gray lines are cell sides. The dashed orange box indicates the jet core region, in which cells have uniform size with a side of 0.75 mm. A contour plot of the instantaneous value of $\dot{\omega}_{pv}$ is included to show that the main reaction zone is entirely found within the jet core. The computational domain starts at the exit of the experimental burner and extends up to 70 D (735 mm). The most upstream part of this domain has a diameter of 10 D that linearly grows up its maximum with 20 D. The employed grid can be seen as an improvement of the coarse grid used in [3], since it preserves similar characteristics of the previous one (e.g., longitudinal and radial cell sizes are the same, however with higher homogeneity here). In [3] a comprehensive grid study has been conducted for reactive and non-reactive flows.

Regarding the boundary conditions, attention is given to the description of the pilot flame. This flame is fueled by a mixture of acetylene and hydrogen at the hydrogen to carbon proportion of the stoichiometric reaction of ethanol. Under adiabatic assumptions, such a boundary condition can be easily represented by adjusting the mixture composition to the stoichiometry. This procedure is previously adopted for simulations presented in [3] and it is commonly employed by diverse research groups (e.g., [7,33,41]) which also used data from the Sydney diluted spray flame burner for modeling validation. Concerning a non-adiabatic formulation, the introduction of the enthalpy transport allows a more accurate representation of the pilot flame. It is important to notice that, despite the same proportion of hydrogen and carbon elements, the presence of the hydroxyl (OH) into the ethanol molecules does not allow the mixture of acetylene and hydrogen to produce the same composition of products. This characteristic becomes clear comparing the stoichiometric reaction of ethanol and the corresponding mixture of acetylene and hydrogen with air

$$C_2H_5OH + 3\,(O_2 + 3.76N_2) \longrightarrow 2CO_2 + 3H_2O + 11.28N_2\,, \tag{18}$$

$$C_2H_2 + 2H_2 + 3.5\,(O_2 + 3.76N_2) \longrightarrow 2CO_2 + 3H_2O + 13.06N_2. \tag{19}$$

Since the main function of a pilot flame consists in the production of hot products to stabilize the main flame, the mixture fraction is adjusted to match the products ratio generated by Equation (19). Namely, the ratio formed by the sum $Y_{CO_2} + Y_{H_2O}$ considering only products. As a result, the following reaction is used to define the mixture composition of the pilot flame for the non-adiabatic simulations performed here

$$0.89C_2H_5OH + 3\,(O_2 + 3.76N_2) \longrightarrow 1.78CO_2 + 2.67H_2O + 11.28N_2 + 0.33O_2\,. \tag{20}$$

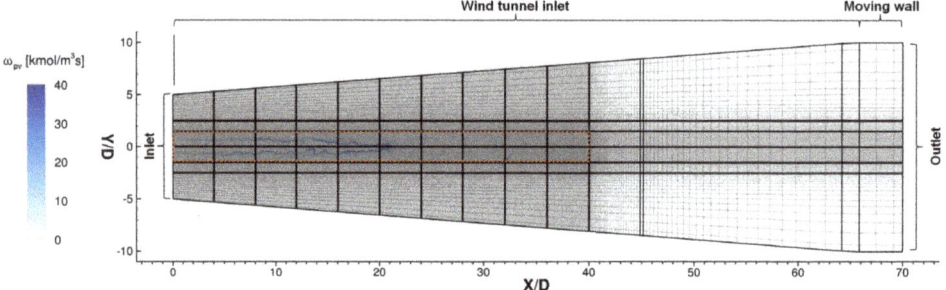

Figure 1. Computational domain. The dashed orange box indicates the jet core.

Once that the absolute enthalpy of the reaction given by Equation (20) does not match with that obtained from Equation (19), the reactants are lightly pre-heated in order to pair up with the desired value. It is important to highlight that with a tabulation method where heat losses are exclusively linked to the evaporation process as in [13,22], such a modeling feature would not be realizable. The relevance of the specification for this boundary condition has been analyzed in [3] for the flame EtF2.

The turbulent boundary condition follows the synthetic turbulence generator proposed by Klein et al. [42]. To account for the pre-vaporized mass of fuel exiting the injection's tube, the mixture fraction receives the explicit value of 0.0165 (corresponding to $\phi_{over} = 0.15$). Effects of the heat transfer throughout the injection tube are considered. For that, the mixture composed by dispersed fuel droplets and air entering at 300 K into the burner exchanges heat with the injections tube wall at 300 K, which results in a mixture exiting the nozzle with 298 K. This arises from our assumption that the mixture of vapor and air with $\phi = 0.15$ and $Re = 38{,}200$ results from the evaporation of liquid in an air stream where both were initially at 300 K. This mixture was then heated up by tube walls at 300 K assuming $Re = 38{,}200$. Considering the remaining boundary conditions for the enthalpy, walls are treated as adiabatic and incoming streams of air are kept at 300 K.

With respect to the boundary conditions of the diluted phase (i.e., droplet size distributions, conditioned velocities on size, injected mass, and injection position), these have been extracted from experimental measurements presented in [31]. An exception occured for the radial component of droplets velocities, for which the procedure suggested by De and Kim [6] was employed here. More details about this procedure are presented in Section 3.1. The initial droplet temperature has been set to 298 K, namely droplets and the pre-vaporized mass of fuel exiting the injection's tube had the same temperature.

Calculations for this research were conducted with the coupled version of the academic software FASTEST (see [3]). Different from previous works, some improvements in numerical methods and algorithms are included. For instance, the ordinary differential equations of the evaporation modeling are integrated in time here with the fifth-order Runge–Kutta scheme of [43].

Simulations are initialized from reactive single-phase computations obtained after 0.055 s, which coincides with one flow-through-time (FTT). The FTT of this configuration corresponds to the time necessary for the gas flow to fill a volume defined from the droplets injection position to 30 D downstream the burner exit, where the last measurement plane is set. Namely, a cylinder with a diameter of 10 D (to account for the burner coflow) and a length of 30 D. Thereafter, parcels start to be injected and a stable number of them is achieved after approximately 1 FTT. Once a stable number of parcels is reached, simulations run for an additional FTT until the averaging begins. Statistics (mean and rms) are collected for 2 FTT for all cases, which amounts circa 35,300 core-hours on 80 CPUs and corresponds to a reduced computational time (RCT) of $\approx 7.11 \times 10^{-8}$ s [RCT = elapsed time × number of CPU/(number of cells × number of iterations)].

To perform the time integration, a second-order, three steps Runge–Kutta scheme (see details in [44]) has been employed. The time-step is maintained constant at 2.4×10^{-6} s. Particles are coupled at each three-time steps. As a result, particles were tracked in a coupling time step of 7.2×10^{-6} s. Despite this discrepancy, the corresponding Courant–Friedrichs–Lewy (CFL) number is kept below 1 respectively for droplets and the gas phase. This coupling strategy alleviates the computational costs of the particle tracking, which amounts in our calculations approximately 50% of the entire computational time.

3. Results

The evaluation of the proposed modeling strategies starts with the establishment of a reference scenario. This corresponds to simulation performed with a fixed exponent β (i.e., $\beta = 0.5$) of Equation (9) and the inclusion of evaporative cooling effects. The value 0.5 resembles the one used in the first investigations of Charlette et al. [28] and is a typical value assumed by different authors (e.g., [9,20,29,44]), therefore it is chosen as a reference here. Altogether three cases are studied, besides the mentioned reference case, one neglecting effects of the evaporative cooling and another considering a parametric variation of the model parameter β while including EC effects. In this last case, the value of β is defined as 0.2.

The influence of the evaporative cooling onto the flow is expressed in terms of the heat necessary to vaporize an amount of liquid (i.e., that associated with the latent heat of vaporization) and the heat-up of droplets. In order to investigate such an influence, EC effects are neglected by removing the terms associated with the latent heat of vaporization and the heat-up of droplets from Equation (4). As a result, only the term including the vapor formation's enthalpy is preserved

$$\overline{S_h} = \sum_{p=1}^{N} \frac{N_p}{V} \left(\dot{m}_{d,p} h_f \right). \tag{21}$$

The source term presented in Equation (21) is necessary for our approach since the energy equation is expressed in terms of the absolute enthalpy. In cases where sensible enthalpy is transported $\overline{S_h} = 0$.

The remainder of this section is divided into two parts. In the first one, simulations results are compared with available experimental data following a validation process. In the second part, analysis of the simulated flames characteristics are conducted in a qualitative fashion.

3.1. Validation of Modeling Strategies

Radial profiles of mean and RMS values of the resolved longitudinal and radial components of droplet velocities are compared with experimental data in Figure 2. As a Cartesian coordinate system is employed the radial direction is represented by the coordinate Y, while the longitudinal by X. Focusing on the first column, one can observe that the proposed strategy is able to recover the mean values of droplets longitudinal velocities reasonably well. Deviations from experimental data are more pronounced downstream of the positions 25 X/D for all cases. Comparing the different simulated cases, further deviations are noticed between positions 5 X/D to 15 X/D. Similar to the observations in [3], it can be clearly noticed that the FSW modeling affects the estimation of mean droplet velocities. Specifically, when $\beta = 0.2$, results better agree with experimental data. Considering this aspect and the fact that such deviations are located in a radial region far from the jet centerline (i.e., Y/D > 0.5), it is possible to state that deviations occur in regions where droplets interact with reaction zones. Nevertheless, due to the strong coupling among the diverse modeling approaches it is not straightforward to draw an exact explanation for it. Influences of velocity fluctuations, thickening factor, and mixture stratification are also supposed to contribute to it, as pointed out throughout the discussions presented in this section. Concerning both cases computed with the same value for the model parameter β, marginal deviations are observed for U_d between them. Close to the jet core (i.e., Y/D < 0.5) both models deliver the same results. Influences are supposed to be caused by the variations of the flame speed caused by heat losses in the flame front as marginal deviations are noticed when Y/D > 0.5 between positions 5 X/D to 15 X/D.

Considering the second column of Figure 2, it is important to highlight that despite the absence of an SGS turbulent dispersion model, droplet velocity fluctuations agree reasonably well with the experimental data. The discrepancies noticed for the positions 10–15 X/D are in great parts assigned to the flame-dispersed phase interaction, which directly influences the droplet tracking. This aspect is supposed to justify the inflection observed when 0.5 < Y/D < 1.0 in the positions 10 X/D and 15 X/D. By comparing the different cases, it is again noticed that the flame wrinkling model also contributes to deviations.

According to the radial profiles of V_d, the experimental values are different from zero at Y/D $\approx$ 0.0, i.e., the jet centerline. As already discussed in [3], the measurements of this velocity component expose a problem on the experimental configuration. For vertical jets, the time averaged radial velocity along the centerline is always zero. However, this is not observed for all comparison planes, including the injection position. This result possibly indicates that the experimental spray is not strictly vertical. Due to this characteristic, the procedure suggested by De and Kim [6] is used here to define the radial velocity of droplets. Namely, the radial velocities of the carrier phase are used to define the starting conditions for droplets. This option justifies the disparity observed in the boundary condition's plane (Figure 2) for V_d. Altogether, simulations results follow a similar trend as the measured data. The existence of non zero values of V_d in the proximity of the jet centerline motivates the addition of a constant offset for all points (for which a zero velocity is attained at the centerline) of the experimental data. Considering them as a reference, a better agreement is noticed for the computed sprays and even better for the case with $\beta = 0.2$. However, because of the anomalies perceived in the experimental measurements (see e.g., spray jet anomaly in [45,46]), further discussions about the radial components of mean droplets' velocity are difficult to be conducted. The same is valid for the last column of Figure 2, where the RMS values of the fluctuations of such velocity components are exhibited. These data are rather used here to give support to the discussions about the different simulated cases.

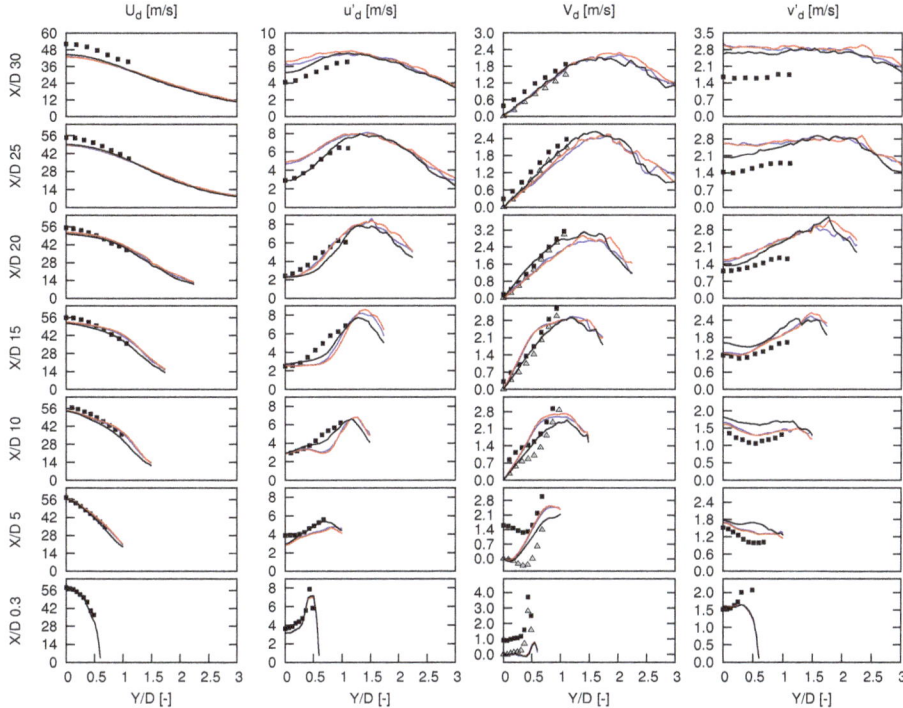

Figure 2. Radial profiles of mean (capital letter) and rms of the fluctuations (lower case letters) of the longitudinal (U_d and u'_d) and radial (V_d and v'_d) components of droplets velocity. Squares-original experimental data, triangles-offset, blue line-evaporative cooling (EC) with $\beta = 0.5$, red line-no EC with $\beta = 0.5$, black line-EC with $\beta = 0.2$.

An important quantity to observe the mass transfer between phases is the liquid volumetric flux (LVF), whose radial profiles at different longitudinal positions are presented in the first column of Figure 3. Both cases computed with $\beta = 0.5$ show lower values for LVF downstream of the position 10 X/D when compared with the case with $\beta = 0.2$. This is a straightforward outcome caused by the increased flame surface wrinkling obtained with $\beta = 0.5$. As the total FSW increases, the consumption flame speed also increases and the overall flame becomes shorter in the longitudinal direction while broader in the radial direction. Consequently, the spray inner core faces a gas flow with higher temperatures than the case computed with $\beta = 0.2$ (see Figure 4). Such higher temperatures increase the evaporation rate and reduce the corresponding LVF. The hot mixture that intensifies the evaporation process is an outcome of the combustion process, which is found in reaction and oxidation zones. As droplets cross a reaction zone, interference with the reaction progress is expected to occur. The amount of droplets interacting with reaction zones has not been quantified in the present work. Nevertheless, studies conducted in [9,16,21] show that FGM tables constructed with premixed flamelets are able to recover the effects of variations of the mixture fraction throughout a flame with good accuracy. The effects of the temperature reduction of the jet inner core and the consequent increase of the simulated LVF are also noticed between both cases computed with $\beta = 0.5$. The inclusion of EC effects indeed decreases the mass transfer between phases as observed by the higher values of LVF downstream of the position 15 X/D. Sacomano Filho et al. [16] indicated by means of flames propagating in droplet mists that the evaporative cooling effects also contribute to the reduction of the laminar flame speed. Bearing this aspect in mind, the increase of LVF and decrease of gas temperature (see Figure 4) when including EC are combined features from the reduction of the inner core temperature and the flame speed. However, the contribution given by the inclusion of EC over the LVF is not so pronounced as that given by the reduced value of the model parameter β. It is important to highlight that the case computed with $\beta = 0.2$ agrees much better with the experimental data than the other cases. Again, this gives an indication of the relevance of the turbulence-flame interaction modeling.

Considering the droplet size distribution, reasonably good agreement with the experimental data can be perceived for the Sauter mean diameter (SMD-D_{32}) in the inner part of the jet (i.e., Y/D < 0.5) and up to 25 X/D. In the farther downstream regions, the SMD becomes slightly smaller than the measured values in all radial positions. A probable cause for it is the penetration of droplets in higher temperature zones, which also agrees with the observed evolution of LVF radial profiles. The interaction of the dispersed phase with reaction zones may justify the deviations noticed for the SMD for radial positions far from the jet centerline (i.e., Y/D > 0.5) from 5 X/D to 20 X/D.

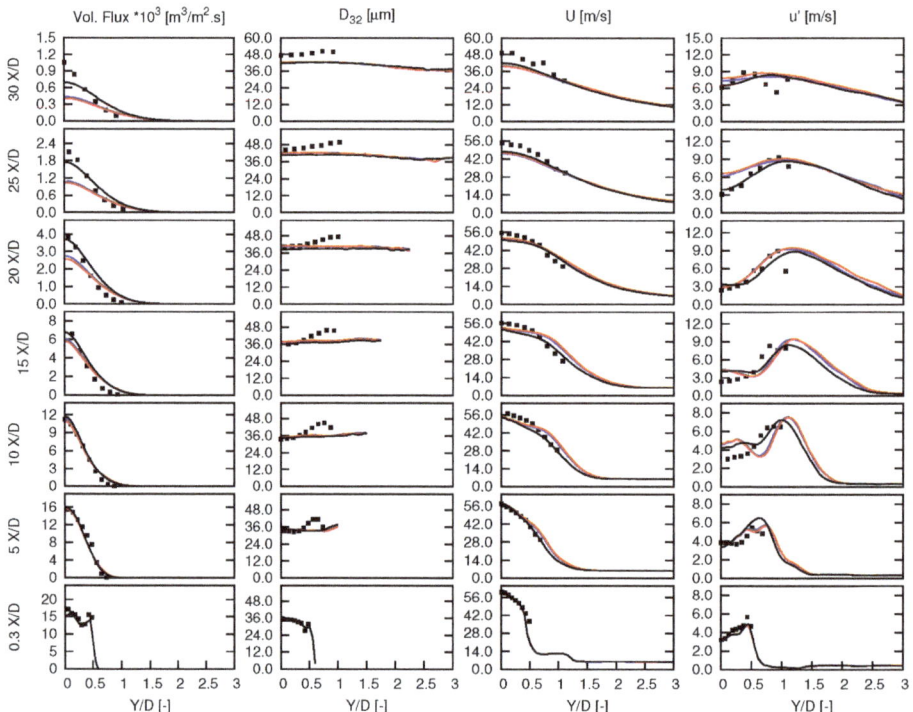

Figure 3. Radial profiles of mean liquid volumetric fluxes, the Sauter mean diameter D_{32}, as well as mean (U) and the rms of the fluctuations (u') of the longitudinal component of the gas velocity. Marks-experiment, blue line-EC with $\beta = 0.5$, red line-no EC with $\beta = 0.5$, black line-EC with $\beta = 0.2$.

In the last two columns in Figure 3, radial profiles of the mean and the RMS values of the resolved longitudinal component of the carrier phase velocity are shown. Experimental data acquired for droplets smaller than 10 µm are used as a reference. In fact, both quantities are not strictly comparable, but the experimental data is used here to guide the subsequent discussions. Many aspects observed for the dispersed phase in Figure 2 are noticed in the gas flow. As for the spray, deviations occur close to reaction zones and in a similar fashion among the different cases. The correct prediction of the flame topology is supposed to contribute to these observed deviations. Concerning the velocity fluctuations, it is important to bear in mind that comparisons performed here do not take into account the contribution of SGS quantities since these are not available. Veynante and Knikker [47] showed that this lack of information does not modify mean but RMS values. As a consequence, the inclusion of these SGS contributions should modify the scenario regarding the rms values.

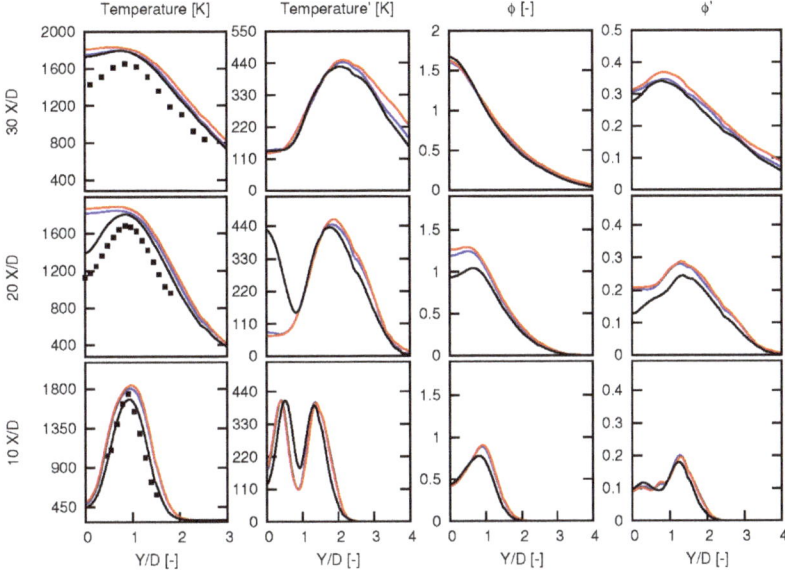

Figure 4. Radial profiles of time averaged and rms of the fluctuation of the equivalence ratio and the temperature of the carrier phase. Marks-experiment, blue line-EC with $\beta = 0.5$, red line-no EC with $\beta = 0.5$, black line-EC with $\beta = 0.2$.

To conclude the validation process, radial profiles of the mean gas temperature are compared with the available experimental data in Figure 4. Results obtained with $\beta = 0.2$ better agree with experiments when compared with those from the remaining cases. Effects of the evaporative cooling are noticed in all positions, but these are more pronounced as the distance from the burner nozzle increases. This is expected since the heat removed from the gas phase increases due to the cumulative impact of the evaporation. However, the effects of FSW modeling are more pronounced than the sole consideration of EC while keeping $\beta = 0.5$. Specifically, deviations between cases computed with $\beta = 0.2$ and $\beta = 0.5$ are more pronounced at 10 X/D and 20 X/D. At 30 X/D, similar results can be seen between cases including EC. This overall behavior regarding FSW modeling is explained considering the flame structures presented in the next section. The FSW modeling is strictly connected with the source term $\dot{\omega}_{pv}$ (see Equation (3)). According to the qualitative contour plots presented in Figure 8, source terms of the progress variable are predominantly found upstream 25 X/D. Hence, the effects of the FSW modeling are more pronounced upstream of this region. Downstream of it, unburned combustion products are mixed together with fuel vapor released from droplets that are able to cross this main reaction zone. The resulting mixture burns further in a diffusion flame mode, which is less affected by the FSW model. Therefore, different approaches for the computation of β deliver similar results at 30 X/D.

Analyzing the temperature results with more details, simulations computed with $\beta = 0.5$ show broader regions with high-temperature values than the results obtained with $\beta = 0.2$ at 10 X/D. In fact, simulations computed with $\beta = 0.2$ agree quite good with the experimental data at this position. Contributions from the inclusion of EC are seen at the region where the temperature peaks, but these are not high. As previously mentioned, effects of the EC are better seen at 20 X/D and 30 X/D throughout the coordinate Y. At 20 X/D, the ascending part of the temperature profile in Y direction for the case in which $\beta = 0.2$ agrees better with experiments than the other cases. The performance of this case at these two positions explain the more accurate prediction of LVF in Figure 3.

3.2. Qualitative Analysis of Simulated Cases

To facilitate the interpretation of the mixture composition, Figure 4 presents time-averaged and RMS values of the equivalence ratio (obtained from the resolved fields of mixture fraction as in [48]). The time averaged values of ϕ point out how strong the FSW model and the inclusion of EC affect the flame structure. Particularly, this variable is connected with the LVF seen in Figure 3, as the vapor released from droplets directly affects the mixture composition. Overall deviations among cases are more pronounced between different definitions of β than the two treatments reserved for the EC. Deviations between both cases computed with $\beta = 0.5$ occur at the position 20 X/D. Observing the time-averaged Lagrangian source terms of mass S_m and enthalpy S_h respectively in Figures 5 and 6, we see that such effect is a direct outcome of the inclusion of EC. Both cases show similar S_m profiles at 20 X/D and upstream this position, while clearly different profiles of S_h. Therefore, the inclusion of EC effects clearly interferes with the mixture composition, as it is a product of higher transported temperatures of the gas phase (see Figure 4) through the higher droplet evaporation rates observed in terms of LVF profiles in Figure 3. With respect to the FSW modeling, the case computed with $\beta = 0.2$ already interferes with the mixture composition at 10 X/D. As seen in Figure 4 and reinforced in Figures 5 and 6, the main contribution to this aspect stems from the region where the reaction zone is found (Figure 8 points out the position of reaction zones by means of the time-averaged $\dot{\omega}_{pv}$), i.e., $0.5 < Y/D < 1.0$. Figures 5 and 6 indicate that small impact of this modeling approach is found in the jet core region, namely $Y/D < 0.5$. As the flame evolves, more interactions with the spray at the jet inner core occur at a downstream position as seen through the Lagrangian source terms at 20 X/D and 30 X/D. The topology of the flame is directly related to this observation. As seen in Figures 7 and 8 upstream of 20 X/D both flames computed with $\beta = 0.5$ depict similar topology. Downstream of this position small flame interactions occur with the spray, whereas droplets that are able to cross the flame tip interact with combustion products coming out of the main flame. The Lagrangian source term profiles indicate that differences between models are also happening at 30 X/D. Nevertheless, these are less intense as no significant impact is noticed in time-averaged profiles of ϕ in Figure 4 at this position.

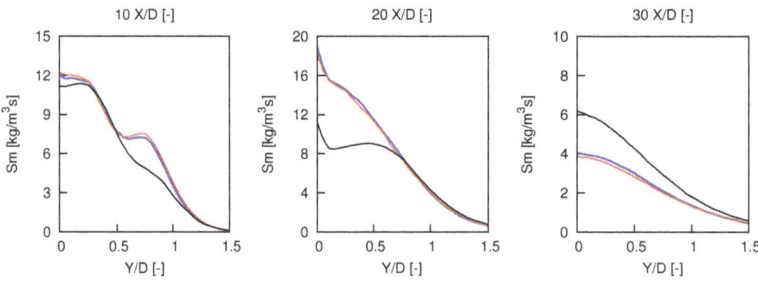

Figure 5. Radial profiles of the Lagrangian source term of mass S_m. Blue line-EC with $\beta = 0.5$, red line-no EC with $\beta = 0.5$, black line-EC with $\beta = 0.2$.

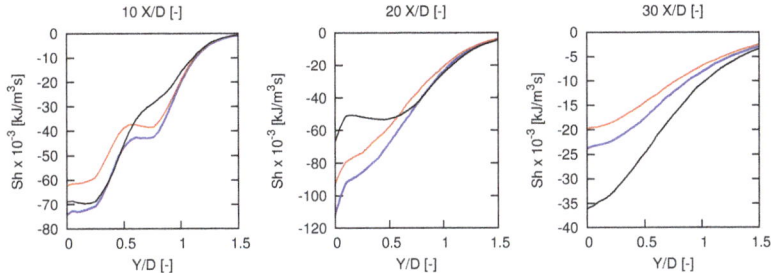

Figure 6. Radial profiles of the Lagrangian source term of enthalpy S_h. Blue line-EC with $\beta = 0.5$, red line-no EC with $\beta = 0.5$, black line-EC with $\beta = 0.2$.

Considering the RMS of ϕ, higher values are found in the region that coincides with the descending part of the temperature profiles. Such an outcome can be seen as a combined effect of the interaction of droplets with high-temperature combustion products and radial flow coming out of the main jet core as seen through the radial profiles of V_d in Figure 2. Deviations among cases are analogous to those found in the time-averaged profiles of the gas temperature. These are significant between FSW modeling approaches at 20 X/D, whereas between EC treatments at 30 X/D.

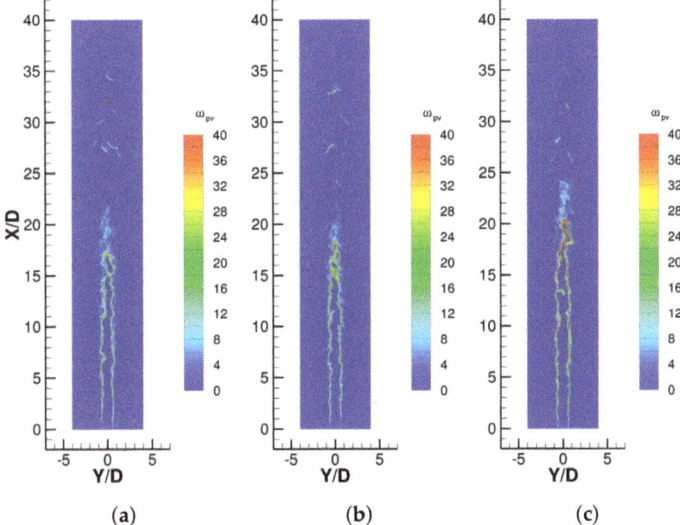

Figure 7. Contour plots of instantaneous values of the source term $\dot{\omega}_{pv}$ [kmol/m³s] in a plane of size 10 D × 40 D at Z/D = 0 for: (a) EC with $\beta = 0.5$; (b) no EC with $\beta = 0.5$; (c) EC with $\beta = 0.2$.

Regarding the RMS of gas temperature, similar trends are noticed among cases, except at 20 X/D. At this position, higher values are found for the case with $\beta = 0.2$ in the proximities of the jet centerline. This characteristic can be explained considering the instantaneous plots of the source term $\dot{\omega}_{pv}$ presented in Figure 7. Different from the cases computed with $\beta = 0.5$, the case with $\beta = 0.2$ deliver a longer flame in which pockets are frequently formed and released close to the flame tip. This occurs close to 20 X/D, where high fluctuations appear for $\beta = 0.2$ in Figure 4.

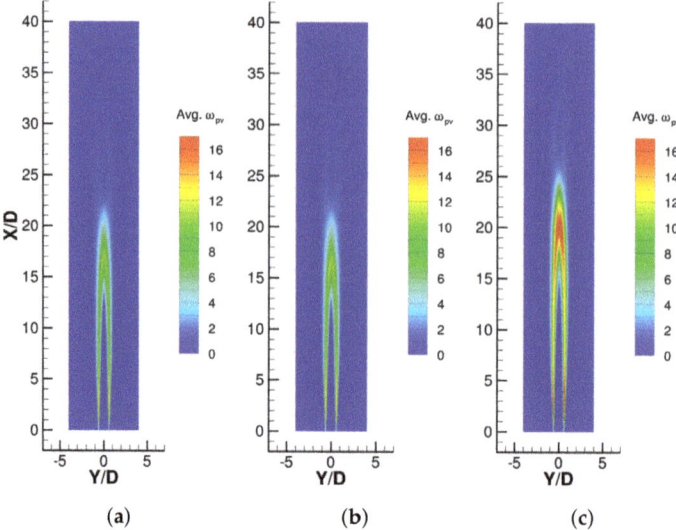

Figure 8. Contour plots of the time averaged source term $\dot{\omega}_{pv}$ [kmol/m^{3}s] in a plane of size 10 D × 40 D at Z/D = 0 for: (**a**) EC with $\beta = 0.5$; (**b**) no EC with $\beta = 0.5$; (**c**) EC with $\beta = 0.2$.

An additional aspect that can be extracted from the contour plots of $\dot{\omega}_{pv}$ presented in Figures 7 and 8 refers to the mixture stratification. Such a feature can be observed in terms of the different intensities of $\dot{\omega}_{pv}$ throughout the flame. The stratification is also noticed in ϕ radial profiles of Figure 4. The variation of the mixture composition within the computational domain is a direct outcome of the evaporation of liquid droplets. Due to the free jet configuration, the amount of released vapor increases as the spray penetrates into the domain. The opposite behavior is noticed in the liquid phase by means of the LVF profiles in Figure 3.

Contour plots of time-averaged values of the gas temperature give a more complete overview of this quantity for the different cases in Figure 9 than the radial profiles shown in Figure 4. The higher penetration length of the flame computed with $\beta = 0.2$ is also observed here considering the penetration of the cold jet core. Additionally, the higher temperatures observed for the case where EC effects are neglected (using the other case computed with $\beta=0.5$ as a reference) are also noticed. The extension of the slice plane used in Figure 9 allows the visualization of another import aspect. Considering the jet centerline (i.e., Y/D = 0.0), as the distance from the nozzle increases, the temperature initially grows from the cold-core level up to a first pick which coincides with the main flame tip position. Downstream of it, the temperature slightly reduces and grows up again for positions above 30 X/D. Particularly, at the region located between 20 X/D and 35 X/D for cases computed with $\beta = 0.5$, whereas between 24 X/D and 38 X/D for the one computed with $\beta = 0.2$, a local cold-core is identified. A similar structure is also perceived, but with much less contrast, in Figure 8 in terms of time-averaged values of the source term $\dot{\omega}_{pv}$. The appearance of it is a result of the combustion of unburnt products mixed with the vapor released from droplets that are able to cross the main reaction zone. In Figure 7 isolated flame structures are found downstream of the main reaction zone. Besides being quite sparse these structures are preferentially located on the jet borders, where the mixture of products and vapor diffuses with the co-flowing air stream. Accordingly, the occurrence of these secondary reactions resemble processes controlled by the flow mixing. This characteristic aligns with the definition of diffusion flames, namely mixing controlled combustion reactions [49]. It is important to highlight that such a multi-regime flame differs from those observed in [10], where the main premixed flame is enveloped by a diffusion-reaction layer. Such an overall behavior is supposed to do not significantly affect the conducted analysis in the present work since the focus has been given to positions upstream of the second cold-core region. Nevertheless, the inclusion of the method recently proposed in [40] to improve the evaporation modeling of droplets interacting with combustion products, as well as the adoption of a multi-regime combustion model (e.g., [50]), may improve the characterization of this post main flame phenomena which may not be reserved to this configuration.

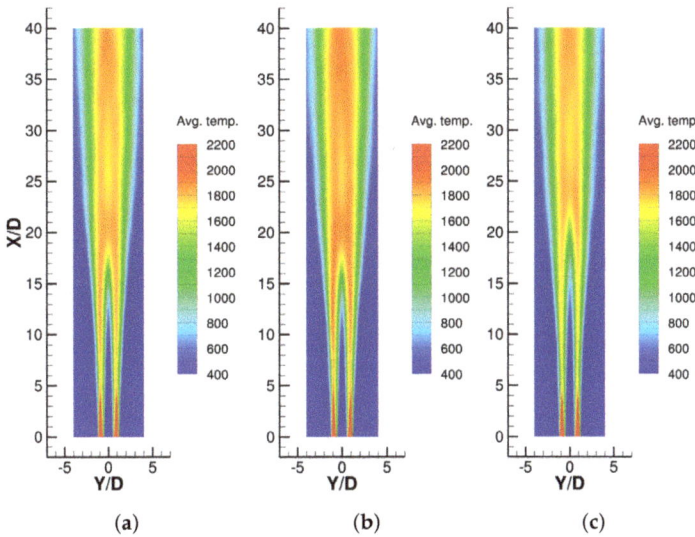

Figure 9. Contour plots of the time averaged values of the gas phase temperature [K] in a plane of size 10 D × 40 D at Z/D = 0 for: (**a**) EC with $\beta = 0.5$; (**b**) no EC with $\beta = 0.5$; (**c**) EC with $\beta = 0.2$.

4. Conclusions

Investigations of the effects caused by the consistent consideration of the evaporative cooling and the turbulence flame interaction demonstrated that both phenomena do interfere not only with the estimation of the flame structure, but also with the prediction of spray properties. As previously pointed out in [3] for a similar flame but with lower Reynolds number, such an outcome reveals the strong interaction among turbulence, flame, and dispersed phase. The necessity of an appropriate computation of each of these phenomena is again reinforced in the present study.

The combination of evaporative cooling effects through the FGM method and the flame surface wrinkling model considering $\beta = 0.2$ demonstrated to have the best performance in the validation process. The corresponding simulated case showed the best overall agreement with available experimental data and could recover the main phenomena observed experimentally. For the validation process, the experimentally measured flame EtF5 of the Sydney diluted spray flame burner [31] was used as a benchmark. The other two cases have been used to allow a parametric study of the influence of the EC effects and FSW modeling. The raised hypothesis in that EC effects could be quite important [16] was accessed by comparing two cases, with and without consideration of EC effects at a rather standard of the FSW model parameter β. Although EC effects were shown to be important for the correct flame prediction, the influence of the turbulence-flame interaction modeling demonstrated to be more pronounced.

Thanks to the phenomena explicitness that LES and ATF approaches allow, the main features observed in the simulation results could be recalled by analyzing further computed quantities. Such a clear interpretation of different phenomena could not be easily done if the proposed modeling extensions to account for heat losses were not consistently done.

Places for improvements are found as well as pointed out throughout the result presentation and discussion. It is important to highlight that the proposed modeling strategy is still limited since part of the underlying phenomena to turbulent spray combustion is not accounted for. Indeed, some effects were not comprehensively studied or even understood to allow to infer on their relevance in the simulations. Two of them are the inclusion of actual mixture composition on droplet evaporation (see e.g., [40]) and effects of differential diffusion in spray flames (see e.g., [16]). Additional effects that may improve the predicting ability of the proposed modeling strategy is, on the one hand, the inclusion of models capable to represent multi-regime combustion, as studied by Knudsen et al. [13] and Hu and Kurose [50], and the incorporation of single droplet burning effects (see [51]) on the other. Nevertheless, it worths to be noticed that the proposed methods following the inclusion of the evaporative cooling effects in the chemistry and the selected FSW modeling converge to an optimum combination of strategies which better approaches the experimental observations. Models able to represent the above-listed phenomena are desired and should improve the capability of the presented module.

Author Contributions: Conceptualization and formal analysis, F.L.S.F.; methodology, F.L.S.F., L.D. and A.H.; software, F.L.S.F. and A.H.; resources, G.C.K.F., and A.S.; writing—original draft preparation, F.L.S.F. and L.D.; supervision, G.C.K.F., and A.S.; funding acquisition, F.L.S.F., G.C.K.F., and A.S.

Funding: This research was funded by São Paulo Research Foundation (FAPESP) grant number 2017/06815-7, as well as by the German Research Council (DFG) grant number TRR 150.

Acknowledgments: We gratefully acknowledge the computational resources by Lichtenberg HPC of the TU Darmstadt and A. Masri to make the experimental data available.

Conflicts of Interest: The authors declare no conflict of interest.

Abbreviations

The following abbreviations are used in this manuscript:

ATF	artificially thickened flame
CFL	Courant–Friedrichs–Lewy
EC	evaporative cooling
FGM	flamelet generated manifolds
FSW	flame surface wrinkling
FTT	flow-through-time
LVF	liquid volumetric flux
RCT	reduced computational time
SGS	sub-grid scale
SMD	Sauter mean diameter

References

1. Jenny, P.; Roekaerts, D.; Beishuizen, N. Modeling of turbulent dilute spray combustion. *Prog. Energy Combust. Sci.* **2012**, *38*, 846–887. [CrossRef]
2. Gutheil, E. *Issues in Computational Studies of Turbulent Spray Combustion*; Springer: Dordrecht, The Netherlands, 2011; pp. 1–39.
3. Sacomano Filho, F.L. Novel Approach Toward the Consistent Simulation of Turbulent Spray Flames Using Tabulated Chemistry. Ph.D. Thesis, Technische Universitaet Darmstadt, Darmstadt, Germany, 2017.
4. Sacomano Filho, F.L.; Chrigui, M.; Sadiki, A.; Janicka, J. Les-based numerical analysis of droplet vaporization process in lean partially premixed turbulent spray flames. *Combust. Sci. Technol.* **2014**, *186*, 435–452. [CrossRef]
5. Chrigui, M.; Masri, A.R.; Sadiki, A.; Janicka, J. Large eddy simulation of a polydisperse ethanol spray flame. *Flow Turbul. Combust.* **2013**, *90*, 813–832. [CrossRef]
6. De, S.; Kim, S.H. Large eddy simulation of dilute reacting sprays: Droplet evaporation and scalar mixing. *Combust. Flame* **2013**, *160*, 2048–2066. [CrossRef]

7. Ukai, S.; Kronenburg, A.; Stein, O.T. Large eddy simulation of dilute acetone spray flames using CMC coupled with tabulated chemistry. *Proc. Combust. Inst.* **2015**, *35*, 1667–1674. [CrossRef]
8. Heye, C.; Raman, V.; Masri, A.R. LES/probability density function approach for the simulation of an ethanol spray flame. *Proc. Combust. Inst.* **2013**, *34*, 1633–1641. [CrossRef]
9. Franzelli, B.; Vié, A.; Boileau, M.; Fiorina, B.; Darabiha, N. Large Eddy Simulation of Swirled Spray Flame Using Detailed and Tabulated Chemical Descriptions. *Flow Turbul. Combust.* **2016**, *98*, 633–661. [CrossRef]
10. Sacomano Filho, F.L.; Kadavelil, J.; Staufer, M.; Sadiki, A.; Janicka, J. Analysis of LES-based combustion models applied to an acetone turbulent spray flame. *Combust. Sci. Technol.* **2018**, *191*, 54–67. [CrossRef]
11. Van Oijen, J.A.; de Goey, L.P.H. Modelling of Premixed Laminar Flames using Flamelet-Generated Manifolds. *Combust. Sci. Technol.* **2000**, *161*, 113–137. [CrossRef]
12. Ketelheun, A.; Kuenne, G.; Janicka, J. Heat transfer modeling in the context of large eddy simulation of premixed combustion with tabulated chemistry. *Flow Turbul. Combust.* **2013**, *91*, 867–893. [CrossRef]
13. Knudsen, E.; Shashank.; Pitsch, H. Modeling partially premixed combustion behavior in multiphase LES. *Combust. Flame* **2015**, *162*, 159–180. [CrossRef]
14. Franzelli, B.; Vié, A.; Ihme, M. Characterizing spray flame-vortex interaction: A spray spectral diagram for extinction. *Combust. Flame* **2016**, *163*, 100–114. [CrossRef]
15. Reveillon, J.; Vervisch, L. Analysis of weakly turbulent dilute-spray flames and spray combustion regimes. *J. Fluid Mech.* **2005**, *537*, 317–347. [CrossRef]
16. Sacomano Filho, F.L.; Speelman, N.; van Oijen, J.A.; de Goey, L.P.H.; Sadiki, A.; Janicka, J. Numerical analyses of laminar flames propagating in droplet mists using detailed and tabulated chemistry. *Combust. Theory Model.* **2018**, *22*, 998–1032. [CrossRef]
17. Boileau, M.; Staffelbach, G.; Cuenot, B.; Poinsot, T.; Bérat, C. LES of an ignition sequence in a gas turbine engine. *Combust. Flame* **2008**, *154*, 2–22. [CrossRef]
18. Tyliszczak, A.; Cavaliere, D.E.; Mastorakos, E. LES/CMC of blow-off in a liquid fueled swirl burner. *Flow Turbul. Combust.* **2014**, *92*, 237–267. [CrossRef]
19. Jones, W.P.; Marquis, A.J.; Noh, D. LES of a methanol spray flame with a stochastic sub-grid model. *Proc. Combust. Inst.* **2015**, *35*, 1685–1691. [CrossRef]
20. Rittler, A.; Proch, F.; Kempf, A.M. LES of the Sydney piloted spray flame series with the PFGM/ATF approach and different sub-filter models. *Combust. Flame* **2015**, *162*, 1575–1598. [CrossRef]
21. Sacomano Filho, F.L.; Kuenne, G.; Chrigui, M.; Sadiki, A.; Janicka, J. A consistent Artificially Thickened Flame approach for spray combustion using LES and the FGM chemistry reduction method: Validation in Lean Partially Pre-Vaporized flames. *Combust. Flame* **2017**, *184*, 68–89. [CrossRef]
22. Olguin, H.; Gutheil, E. Influence of evaporation on spray flamelet structures. *Combust. Flame* **2014**, *161*, 987–996. [CrossRef]
23. Hu, Y.; Olguin, H.; Gutheil, E. A spray flamelet/progress variable approach combined with a transported joint PDF model for turbulent spray flames. *Combust. Theory Model.* **2017**, *21*, 575–602. [CrossRef]
24. Luo, K.; Fan, J.; Cen, K. New spray flamelet equations considering evaporation effects in the mixture fraction space. *Fuel* **2013**, *103*, 1154–1157. [CrossRef]
25. Franzelli, B.; Fiorina, B.; Darabiha, N.; Paris, E.C. A tabulated chemistry model dedicated to spray combustion Mixture fraction Yz. *Eng. Sci.* **2012**. [CrossRef]
26. O'Rourke, P.J.; Bracco, F.V. Two scaling transformations for the numerical computation of multidimensional unsteady laminar flames. *J. Comput. Phys.* **1979**, *33*, 185–203. [CrossRef]
27. Colin, O.; Ducros, F.; Veynante, D.; Poinsot, T. A thickened flame model for large eddy simulations of turbulent premixed combustion. *Phys. Fluids* **2000**, *12*, 1843–1863. [CrossRef]
28. Charlette, F.; Meneveau, C.; Veynante, D. A power-law flame wrinkling model for LES of premixed turbulent combustion Part I: Non-Dynamic Formulation and Initial Tests. *Combust. Flame* **2002**, *131*, 159–180. [CrossRef]
29. Kuenne, G.; Seffrin, F.; Fuest, F.; Stahler, T.; Ketelheun, A.; Geyer, D.; Janicka, J.; Dreizler, A. Experimental and numerical analysis of a lean premixed stratified burner using 1D Raman/Rayleigh scattering and large eddy simulation. *Combust. Flame* **2012**, *159*, 2669–2689. [CrossRef]
30. Charlette, F.; Meneveau, C.; Veynante, D. A power-law flame wrinkling model for LES of premixed turbulent combustion Part II: Dynamic formulation. *Combust. Flame* **2002**, *131*, 181–197. [CrossRef]

31. Gounder, J.D.; Kourmatzis, A.; Masri, A.R. Turbulent piloted dilute spray flames: Flow fields and droplet dynamics. *Combust. Flame* **2012**, *159*, 3372–3397. [CrossRef]
32. Marinov, N.M. A detailed chemical kinetic model for high temperature ethanol oxidation. *Int. J. Chem. Kinet.* **2002**, *31*, 183–220. [CrossRef]
33. Chrigui, M.; Gounder, J.; Sadiki, A.; Masri, A.R.; Janicka, J. Partially premixed reacting acetone spray using LES and FGM tabulated chemistry. *Combust. Flame* **2012**, *159*, 2718–2741. [CrossRef]
34. Germano, M.; Piomelli, U.; Moin, P.; Cabot, W.H. A dynamic subgrid-scale eddy viscosity model. *Phys. Fluids A* **1991**, *3*, 1760–1765. [CrossRef]
35. Aschmoneit, K. Numerische Beschreibung Technischer Verbrennungssysteme. Ph.D. Thesis, Technische Universitaet Darmstadt, Darmstadt, Germany, 2013.
36. Durand, L.; Polifke, W. Implementation of the Thickened Flame Model for Large Eddy Simulation of Turbulent Premixed Combustion in a Commercial Solver. In Proceedings of the ASME Turbo Expo 2007: Power for Land, Sea, and Air, Montreal, QC, Canada, 14–17 May 2007.
37. Wang, G.; Boileau, M.; Veynante, D. Implementation of a dynamic thickened flame model for large eddy simulations of turbulent premixed combustion. *Combust. Flame* **2011**, *158*, 2199–2213. [CrossRef]
38. Ketelheun, A.; Olbricht, C.; Hahn, F.; Janicka, J. Premixed Generated Manifolds for the Computation of Technical Combustion Systems. In Proceedings of the ASME Turbo Expo 2009: Power for Land, Sea, and Air, Orlando, FL, USA, 8–12 June 2009; pp. 695–705.
39. Miller, R.S.; Harstad, K.; Bellan, J. Evaluation of equilibrium and non-equilibrium evaporation models for many-droplet gas-liquid flow simulations. *Int. J. Multiph. Flow* **1998**, *24*, 1025–1055. [CrossRef]
40. Sacomano Filho, F.L.; Krieger Filho, G.C.; van Oijen, J.A.; Sadiki, A.; Janicka, J. A novel strategy to accurately represent the carrier gas properties of droplets evaporating in a combustion environment. *Int. J. Heat Mass Transf.* **2019**, *137*, 1141–1153. [CrossRef]
41. De, S.; Lakshmisha, K.N.; Bilger, R.W. Modeling of nonreacting and reacting turbulent spray jets using a fully stochastic separated flow approach. *Combust. Flame* **2011**, *158*, 1992–2008. [CrossRef]
42. Klein, M.; Sadiki, A.; Janicka, J. A digital filter based generation of inflow data for spatially developing direct numerical or large eddy simulations. *J. Comput. Phys.* **2003**, *186*, 652–665. [CrossRef]
43. Cash, J.R.; Karp, A.H. A variable order Runge-Kutta method for initial value problems with rapidly varying right-hand sides. *ACM Trans. Math. Softw.* **2002**, *16*, 201–222. [CrossRef]
44. Kuenne, G.; Ketelheun, A.; Janicka, J. LES modeling of premixed combustion using a thickened flame approach coupled with FGM tabulated chemistry. *Combust. Flame* **2011**, *158*, 1750–1767. [CrossRef]
45. Bart, M.; Roekaerts, D.; Sadiki, A. *Experiments and Numerical Simulations of Diluted Spray Turbulent Combustion*, 1st ed.; Springer: Dordrecht, The Netherlands, 2011; Volume 17, pp. 140–143.
46. Gounder, J.D. An Experimental Investigation of Non-Reacting and Reacting Spray Jets. Ph.D. Thesis, The University of Sydney, Sydney, Australia, 2009.
47. Veynante, D.; Knikker, R. Comparison between LES results and experimental data in reacting flows. *J. Turbul.* **2006**, *7*, N35. [CrossRef]
48. Maschinenbau, V.F. *Large Eddy Simulation of Premixed Combustion Using Artificial Flame Thickening Coupled with Tabulated Chemistry an der Technischen Universität Darmstadt Dissertation*, 1st ed.; Optimus Verlag: Göttingen, Germany, 2012; p. 282.
49. Poinsot, T.; Veynante, D. *Theoretical and Numerical Combustion*, 1st ed.; R. T. Edwards: Philadelphia, PA, USA, 2001; p. 473.
50. Hu, Y.; Kurose, R. Partially premixed flamelet in LES of acetone spray flames. *Proc. Combust. Inst.* **2018**. [CrossRef]
51. Bojko, B.T.; DesJardin, P.E. On the development and application of a droplet flamelet-generated manifold for use in two-phase turbulent combustion simulations. *Combust. Flame* **2017**, *183*, 50–65. [CrossRef]

© 2019 by the authors. Licensee MDPI, Basel, Switzerland. This article is an open access article distributed under the terms and conditions of the Creative Commons Attribution (CC BY) license (http://creativecommons.org/licenses/by/4.0/).

Article

Modelling of Self-Ignition in Spark-Ignition Engine Using Reduced Chemical Kinetics for Gasoline Surrogates

Ahmed Faraz Khan [1], Philip John Roberts [2] and Alexey A. Burluka [3,*]

1. Jaguar Land Rover, W/10/3, Abbey Road, Whitley, Coventry CV3 4LF, UK
2. HORIBA MIRA Nuneaton, Warwickshire CV10 0TU, UK
3. Faculty of Engineering and Environment, Northumbria University, Newcastle-upon-Tyne NE1 8ST, UK
* Correspondence: alexey.burluka@northumbria.ac.uk

Received: 28 May 2019; Accepted: 14 August 2019; Published: 17 August 2019

Abstract: A numerical and experimental investigation in to the role of gasoline surrogates and their reduced chemical kinetic mechanisms in spark ignition (SI) engine knocking has been carried out. In order to predict autoignition of gasoline in a spark ignition engine three reduced chemical kinetic mechanisms have been coupled with quasi-dimensional thermodynamic modelling approach. The modelling was supported by measurements of the knocking tendencies of three fuels of very different compositions yet an equivalent Research Octane Number (RON) of 90 (ULG90, PRF90 and 71.5% by volume toluene blended with n-heptane) as well as iso-octane. The experimental knock onsets provided a benchmark for the chemical kinetic predictions of autoignition and also highlighted the limitations of characterisation of the knock resistance of a gasoline in terms of the Research and Motoring octane numbers and the role of these parameters in surrogate formulation. Two approaches used to optimise the surrogate composition have been discussed and possible surrogates for ULG90 have been formulated and numerically studied. A discussion has also been made on the various surrogates from the literature which have been tested in shock tube and rapid compression machines for their autoignition times and are a source of chemical kinetic mechanism validation. The differences in the knock onsets of the tested fuels have been explained by modelling their reactivity using semi-detailed chemical kinetics. Through this work, the weaknesses and challenges of autoignition modelling in SI engines through gasoline surrogate chemical kinetics have been highlighted. Adequacy of a surrogate in simulating the autoignition behaviour of gasoline has also been investigated as it is more important for the surrogate to have the same reactivity as the gasoline at all engine relevant $p - T$ conditions than having the same RON and Motored Octane Number (MON).

Keywords: autoignition modelling; reduced chemical kinetics; gasoline surrogates; engine knock

1. Introduction

The knock phenomenon limits the compression ratio and the thermodynamic efficiency of a spark ignition (SI) engine. At some operating conditions, avoiding it will force the spark advance away from the maximum brake torque point further lowering engine performance. If knock occurs, it can be severely damaging to the engine components. Knock, i.e., onset of strong pressure oscillations, has been known to be caused by the autoignition of the end gas ahead of the propagating flame which is controlled by chemical kinetics. Therefore, numerical modelling of autoignition may be used to predict knock onset. This has been demonstrated in various studies by modelling the autoignition chemistry of a simpler gasoline surrogate such as primary reference fuels (PRFs), i.e., a mixture of iso-octane and n-heptane [1–3].

Chemical kinetic mechanisms are routinely validated against shock tube and rapid compression machine measurements of the ignition delay times (τ_{ign}) of various gasoline surrogate fuels. The conditions for these laboratory measurements are similar to those which prevail before ignition in homogeneous charge compression ignition (HCCI) or controlled autoignition (CAI) engine, and the recent kinetic mechanisms, e.g., [4–6], perform remarkably well in these regimes. Application of such gasoline surrogate mechanisms have been seen in various studies demonstrating autoignition predictions in HCCI and CAI engines, [4,7]. The question arises of whether the reduced chemical mechanisms will perform equally well in predicting autoignition in SI engines. This work attempts to address this subject.

2. Practical Gasoline Surrogates

The simplest surrogate, a primary reference fuel (PRF), i.e., mixture of iso-octane and n-heptane, is used in the well known Research and Motored Octane Number (RON and MON) tests to quantify the knock resistance of a gasoline by matching knock intensities in a standard engine. Compared to the RON test, in the MON test the end gas temperature is higher at the same pressures and therefore, unless gasoline exhibits a negative temperature coefficient (NTC) phase at those temperatures, the MON value tends to be lower than RON. The sensitivity S, i.e., the difference of octane numbers ($S = RON - MON$) is a measure of fuel's response to varying pressures and temperatures. PRF's have zero sensitivity by definition. However, gasoline, owing to its complex composition, is likely to exhibit autoignition behaviour different from that of the PRF which matches its RON or the MON value, when the pressure and temperature history of the end gas does not match the standard RON and MON tests. Blends of iso-octane, n-heptane with toluene (Toluene Reference Fuels, TRFs) have an advantage of having non-zero sensitivity. Binary blends of toluene with iso-octane and n-heptane have also been studied in order to isolate their cross-oxidation chemistry [8], however such blends have found little application as gasoline surrogates.

Use of iso-octane and n-heptane as components of a surrogate is hardly avoidable as they represent linear and branched alkanes, major gasoline components. Besides, the chemistry leading to their auto-ignition is relatively well understood. However, gasoline does not consists only of alkanes. Thus, EN228, the European standard for gasoline specifies aromatic content of up to 35% by volume, it also allows 5% oxygenates by volume. Therefore, it seems natural to seek surrogates going beyond PRF's and TRF's which contain compounds approximating various families of hydrocarbons present in the gasoline [9].

Table 1 presents a list of such gasoline surrogates for which the auto-igntion delay times were experimentally determined using rapid compression machines (RCM), shock tubes (ST) and HCCI engines. The work of Gauthier et al. [10] presents a fairly comprehensive shock tube study of two TRFs for $p-T$ conditions of 12–25 and 45–60 atm at 850–1280 K. Two TRF's were proposed to approximate a standard research gasoline, RD387, with an anti knock index (AKI) of 87. The measurements demonstrated a good similarity between the surrogates and the gasoline. However the RON and MON of these surrogates were later determined by Knop et al. [11] and the AKI of the surrogates were found to be lower than 87, see Table 1. But, even more importantly, the sensitivity S of the two TRFs was found to be much lower than the usual gasoline range of S of 8 to 12 points.

While shock tube measurements offer an important benchmark for validation of the chemical kinetic mechanisms, they are difficult to perform at temperatures below 850 K and this makes validation of chemical kinetic mechanisms at engine conditions difficult. At these temperatures, many individual components of gasolines have ignition delays decreasing with temperature, the phenomenon known as negative temperature coefficient (NTC) behaviour. Work of Mehl et al. [12] demonstrated that the sensitivity of a surrogate can be correlated to the slope of the NTC region, $dlog(\tau_{ign})/dT$, while the values of the auto-ignition delay time τ_{ign} in the NTC region depend on the AKI of the surrogate. These two correlations and knowledge of the gasoline composition provides constraints for the aromatic and olefinic content of the surrogate which are key to achieving a realistic NTC behaviour and thus sensitivity. The PRF content was then varied to achieve the correct H/C ratio and the octane numbers. This approach was applied to another RD387 gasoline [12], for which a 4-component surrogate (Table 1, last entry) was proposed. Work of Kukkadapu et al. [13] further investigated one of the TRF formulation proposed in [10]; this surrogate is referred to as Gauthier TRF-A (iso-octane 63%, n-heptane 17%, toluene 20% by volume). Comparison of igntion delays for Gauthier TRF-A and the 4-component surrogate proposed by Mehl et al. [12], measured in an RCM, Kukkadapu et al. [13] revealed that the correlations of Mehl et al. [12] overestimate the fuel sensitivity thereby indicating a weakness of an empirical prediction of octane numbers.

Table 1. Volumetric composition of various gasoline surrogate blends the igintion delays of which have been measured.

Iso-Octane	n-Heptane	Toluene	Ethanol	Olefin	RON/MON	Reference
% by Volume						
63	17	20	-	-	86.6/84.2 *	Gauthier et al. [10]
69	17	14	-	-	85.7/84.6 *	Gauthier et al. [10]
62	18	-	20	-	95.1/89.5	Fikri et al. [14]
25	20	45.0	-	10 (DIB **)	94.6/85	Fikri et al. [14]
37.8	10.2	12	40	-	-	Cancino et al. [15]
30	22	25	10	13 (DIB)	-	Cancino et al. [15]
57	16	23	-	4 (C_5H_{10}−2)	91/83 est.	Mehl et al. [12]

* determined by Knop et al. [11]; ** di-iso-butylene (C_8H_{16}); est. estimated values.

Ethanol addition to gasoline is now common mainly due to legislative impetus. Ethanol acts as an octane improver and is commonly blended in amounts of up to 10% by volume. Works of Fikri et al. [14] and Cancino et al. [15] performed shock tube measurements of four multicomponent gasoline surrogates including ethanol, see Table 1.

The EN 228 gasoline standard specifies a maximum olefin content of 18% vol. Most European gasolines have olefin content between 5% and 9% vol, mostly branched rather than straight or cyclic compounds [9]. Oxidation characteristics of some olefins, such as 1-hexene, cyclohexene and 1-pentene have been studied, however, other very common ones, e.g., 2-methyl-2-butene are little studied. It is because of this lack of understanding of common gasoline olefins that proposed chemical kinetic schemes differ significantly in olefine oxidation pathways. For autoignition simulations, the choice of olefin is crucial for producing correct ignition delays while also matching the H/C ratio. Surrogates tabulated in Table 1 have therefore only limited capability to emulate gasoline in SI engine conditions.

3. Chemical Kinetics Schemes for Gasoline Surrogates

An earlier study [16] compared autoignition predictions in an adiabatic homogeneous reactor and an HCCI engine using 8 chemical kinetics mechanisms of different sizes. However, these mechanisms can describe only a limited amount of gasoline compounds, this is why this work turns to more comprehensive, semi-detailed kinetic schemes referred to as Andrae's [4], Golovitchev's [5], and Reitz' [6], mechanisms. For individual substances, all these three mechanisms are capable of producing ignition delay times in fairly good agreement to experiments, see Figure 1. It may be inferred from this Figure that biggest discrepancy between experiments and models arises at high pressures and low, 600–850 K, temperatures. The observed difference would amount to a substantial deviation in terms of crank angle timing, e.g., at 1500 rpm, 1 ms equals 9° of the crank rotation. The integrated error along the complete $p - T$ history of the end gas may therefore result in a substantial difference between the predicted and observed autoignition onsets.

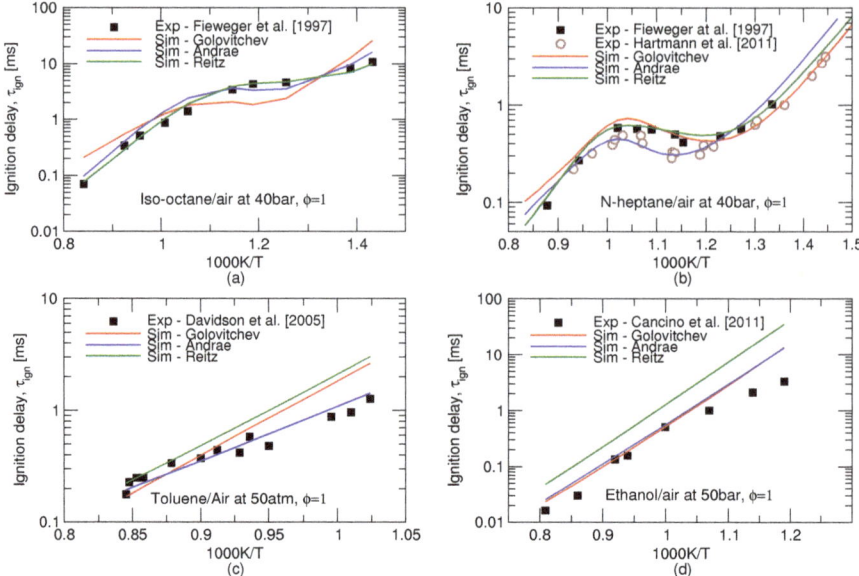

Figure 1. Ignition delay time predictions (lines) of four individual key gasoline surrogate components: (**a**) iso-octane, (**b**) n-heptane, (**c**) toluene (**d**) ethanol. Symbols show shock tube measurements of Fieweger et al. [17], Hartmann et al. [18], Davidson et al. [19] and Cancino et al. [15].

Performance of the three reduced mechanisms studied in this work for the ignition delay time predictions for the gasoline surrogates remains similar to the one for the individual icomponents. Ignition delay times for the stoichiometric mixtures of two surrogates at high pressures is presented in Figure 2. The over-prediction of the NTC behaviour of iso-octane by Golovitchev model as seen in Figure 1a is inheriteded in the surrogate simulations as well and, as a consequence, considerably lower ignition delay times are produced for both surrogates. It can be seen that Andrae's mechanism outperforms the others at low, 600–850 K, temperatures crucial for the prediction of autoignition in SI engines.

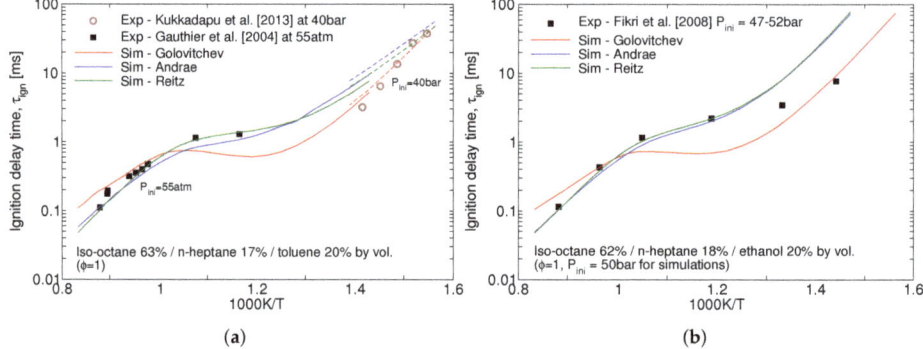

Figure 2. Predicted vs. measured ignition delay times of the (**a**) Gauthier TRF-A and (**b**) an ethanol-containing gasoline surrogate by Fikri et al. [14].

The chemical kinetic calculations presented in this work have been done using a solver routine written in Fortran as part of the library of 1-D engine modelling code at Leeds University, refered to as LUSIE (Leeds University Spark Ignition Engine). For chemical kinetics, the solution of the stiff-type differential equations is carried out by using an implementation of the modified extended backward differentiation formulas developed by Cash [20]. The chemistry solver has been coupled with the engine simulation package in which the main combustion event is modelled using a multi-zone thermodymanic approach to flame propagation [21].

4. Gasoline Surrogate Formulation

Optimal surrogate formulation depends on its application. A large number of physical and chemical properties of the target fuel can be targeted in the surrogate, among those are the distillation curve, RON, MON, stoichiometric air-to-fuel ratio, molecular weight, thermal conductivity and laminar burning velocity. At the same time, the number of components in gasoline surrogate is limited by the availability of chemical kinetic data and complexity of blending rules. Since gasoline constituents belong to one of the five main classes it seems natural to seek surrogate with the same number of components. One mathematical constraint which must always be met while determining the composition of a surrogate is that the sum of the mole or volume fractions of its constituents must be one. This means that for a n component surrogate, $n-1$ properties can be used as constraints to optimise the surrogate composition. For a correct prediction of the cumulative heat release the stoichiometric air to fuel ratio (A/F_s) and the calorific value of the gasoline must also be matched. The correct H/C/O atomic proportions and the molar mass, M will automatically produce the desired A/F_s.

For the quasi-dimensional combustion modelling in SI engine coupled with chemical kinetic modelling of surrogate autoignition in the end gas, the H/C and O/C ratios are crucial. Therefore, these constraints are used in the determination of the surrogate composition:

$$\frac{\sum_{i=1}^{n} x_i H_i}{\sum_{i=1}^{n} x_i C_i} = H/C \qquad \frac{\sum_{i=1}^{n} x_i O_i}{\sum_{i=1}^{n} x_i C_i} = O/C \qquad (1)$$

Reproducing the autoignition behaviour of the gasoline is the biggest challenge in surrogate formulation. The capablity of the surrogate to represent the anti-knock properties of the petrol is only partially dependent on the used model for octane number. A *true* surrogate for gasoline autoignition will match ignition delay of the gasoline at all conditions. It will show the emergence of similar ignition precursors at similar rates to that of the gasoline and therefore it will have, not only the same RON and MON as the gasoline, but similar octane index no matter which engine the two are compared in. Matching just the RON and MON of the surrogate with that of the gasoline does not guarantee that the surrogate will reproduce the autoignition behaviour of the gasoline universally in all engines. To make matters worse, the empirical/theoretical octane number models are far from perfect.

Detailed composition-based octane number models such as [22] account for the non-linear blending interactions of surrogate constituents. The model developed in [22] accounts for the paraffin-olefin and paraffin-naphthene interactions. The non-linear octane blending between ethanol and other gasoline constituents is a subject of on-going research [23] but as yet no well tested octane number model has emerged. Pera and Knop [9] advocated the use of a linear-by-moles additivity rule for the TRF blends shown to perform better than the non-linear model of Morgan et al. [24] or the composition-based octane model of Ghosh et al. [22]. Pera and Knop [9] proposed an improvement to the linear-by-moles expression for TRFs by suggesting blend octane numbers for toluene (RON 116/MON 101.8) and demonstrated that their expression yielded the lowest absolute errors in comparison to 7 other octane number models [11]. Their approach has been found to produce appreciable octane numbers in the present work and due to its accuracy and simplicity it has been adopted in this work for the calculation of TRF octane numbers as well as surrogates containing olefin:

$$\sum_{i=1}^{n} ON_i x_i = ON \qquad (2)$$

Equations (1) and (2) and the unity sum of molar fractions provide 5 constraints for the determination of the so-called properties-based surrogate. Alternative to this would be a composition-based surrogate formulated by representing major constituents of gasoline by a surrogate of that particular family. The resulting surrogate will be expected to have different properties from that of the target gasoline as the surrogate components do not correctly represent all of the substances in that family. However, one of the findings of this work is that a composition-based surrogate whose composition represents accurately fractions of the actual gasoline may perform superior in replicating the autoignition behaviour as compared with a purely properties based surrogate which may contain unrealistic amounts of aromatics and oxygenates.

5. Supporting Experiments

Experiments were performed in a single cylinder optical engine (LUPOE2-D) at Leeds University with a disc shaped combustion chamber, the details of the engine and the experimental methods has been presented in [25]. Iso-octane, as well as three fuels of very different compositions, commercial unleaded petrol ULG90, PRF and TRF, with the same RON of 90 were tested in the knocking regime to assess the differences in their auto-ignition behaviour. For all fuels stoichiometric mixtures were tested. TRF90 was blended with toluene and n-heptane only as most of the octane quality of ULG90 came from its branched paraffin content.

For autoignition modelling, cycles with similar pressure traces were chosen for the different fuels to allow a comparison of the knocking tendencies of different fuels at similar conditions, Figure 3. The subject of autoignition modelling with predictive combustion for the full range of slow and fast burning cycles has been covered in [1–3]. For this study, because fuel composition has only a small effect on the heat loss during compression phase, matching the cycle pressure effectively means very similar end-gas temperatures will result.

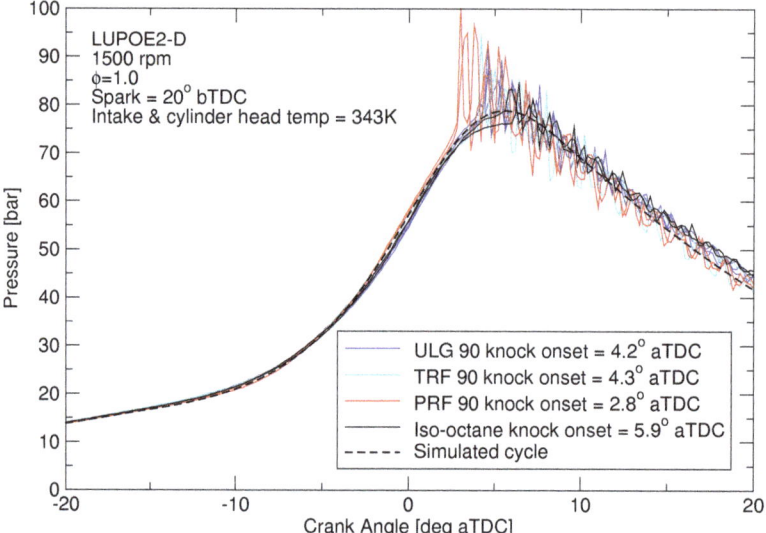

Figure 3. The knocking experimental cycles for the four different fuels which have pressures similar to the pre-knock pressure of ULG90.

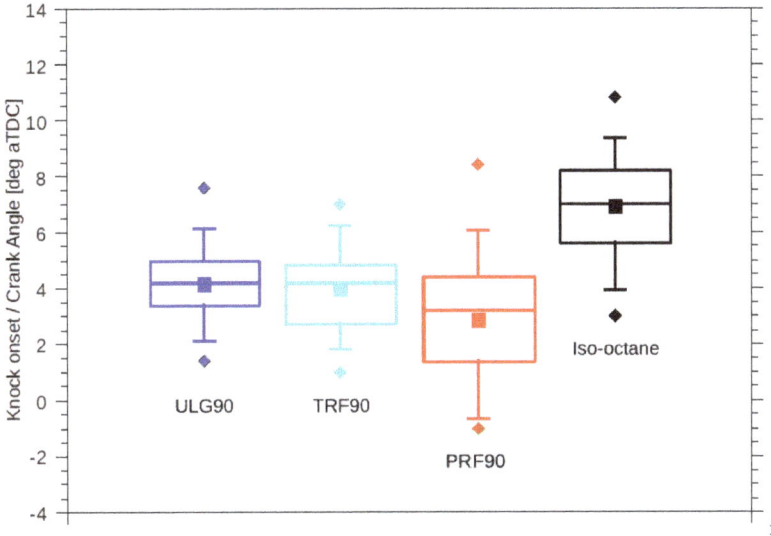

Figure 4. Statistical variation in measured knock onsets for 100 combustion cycles of the four fuels tested. Squares: mean knock onset, diamonds: earliest and latest knock onsets, box: 25/75 percentile, whiskers: 5/95 percentile.

Difference in the reactivity of the two fuels was expected to manifest as different knock onsets regardless of the equivalent RON. However, TRF90 resulted in similar knock onset as ULG90 unlike PRF90 (Figure 4). The effective octane rating of ULG90 and TRF90 appears to be superior than the corresponding PRF of ON 90.

6. Results and Discussion

The relationship between the end gas thermodynamic state and the autoignitive tendency of a fuel can be depicted as ignition delay time (τ_{ign}) contours on a $p - T$ carpet plot as shown in Figures 5–7.

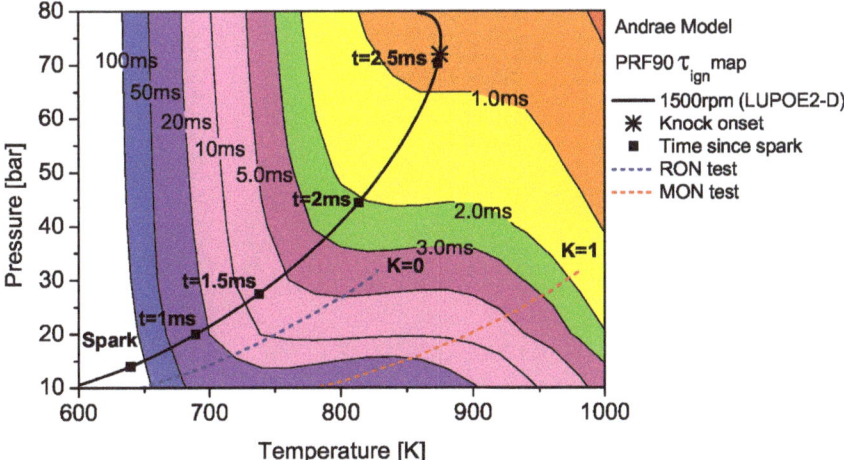

Figure 5. The τ_{ign} contour map of PRF90 as predicted by the Andrae model superimposed with the unburned zone $p - T$ history of LUPOE2-D and typical RON/MON tests.

The end gas thermodynamic path in LUPOE2-D as well as the RON and MON tests are superimposed on the τ_{ign} maps for PRF90 and TRF90; the latter calculated using Andrae's model. The differences in fuel reactivity are shown by the shape of the contours and the extent to which the regions of different τ_{ign} span. In the case of PRF90, the LUPOE2-D $p - T$ trajectory follows a path before the start of NTC phase where the regions of shorter τ_{ign} are met earlier. Combustion phasing with respect to the top dead centre (TDC) is such that the regions of longer τ_{ign} are passed relatively slowly as the pressure increase is initially slower. This phase of combustion corresponds to the period of the initial acceleration of the flame after spark until it reaches a steady turbulent burning velocity after which it starts to decelerate approaching the wall [26].

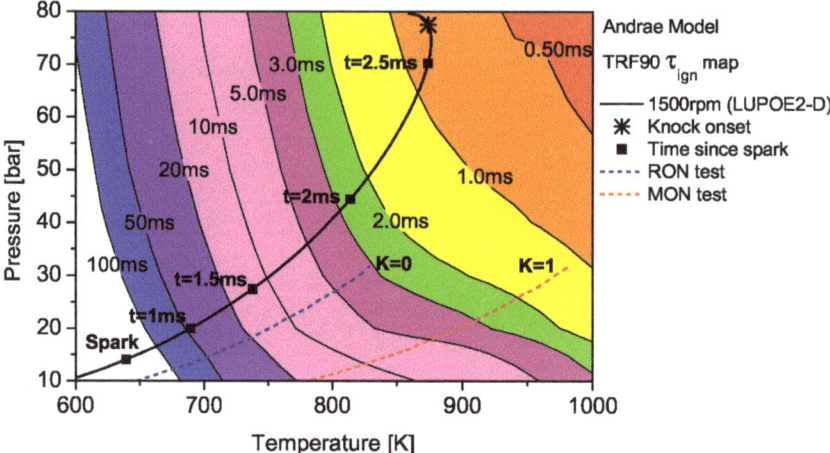

Figure 6. The τ_{ign} contour map of TRF90 as predicted by the Andrae model superimposed by the unburned zone $p - T$ history of LUPOE2-D and typical RON/MON tests.

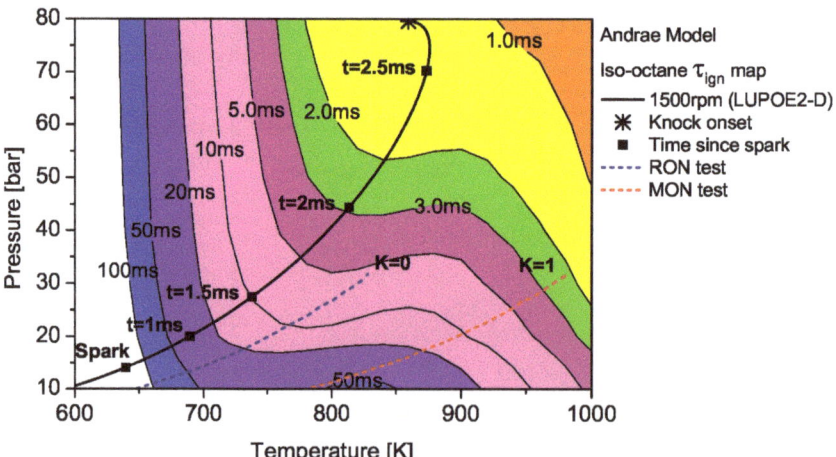

Figure 7. The τ_{ign} contour map of iso-octane as predicted by the Andrae model superimposed by the unburned zone $p - T$ history of LUPOE2-D and typical RON/MON tests.

The integration of the individual τ_{ign} along an engine's $p - T$ trajectory using the Livengood-Wu integral suggests that it is the short ignition delay region towards the later stages of combustion when the $p - T$ are at their highest, which contribute the most in producing ignition. Comparison of the τ_{ign}

contours for PRF90, TRF90 and iso-octane shows that the regions of short τ_{ign} are situated at lower $p - T$ domain for PRF90 and a significant portion of the LUPOE2-D trajectory lies in these critical regions bringing the knock onset to an earliest value of 2.8° CA after TDC (aTDC) among the four fuels studied. Iso-octane shows the latest knock onset as the $p - T$ trajectory stays within fairly long τ_{ign} regions, Figure 7.

The end gas thermodynamic path in a RON test passes the same amount of time in regions of very similar τ_{ign} for both PRF90 and TRF90, thus giving them the same RON values. This means that PRF90 can be regarded as a surrogate for TRF90 and ULG90 at only such the unburned zone $p - T$ conditions as encountered in the RON test. Similarly a PRF with octane number of 84.7 may be regarded as a surrogate for ULG90 at $p - T$ conditions of MON test. However, a single multi-component blend can be used to reproduce the autoignition behaviour of a gasoline at both RON and MON test $p - T$ conditions. Such a blend which has been matched to the RON/MON of a gasoline will have the same reactivity for only a narrow region of the $p - T$ landscape. It is therefore tempting to consider such a blend as a surrogate at all $p - T$ conditions. The fact that most modern SI engines operate with temperature and pressure history very similar to LUPOE2-D or at even higher pressures resulting from turbocharging, RON and MON are not sufficient to characterise auto-ignition of gasoline in such engines. Moreover, weaknesses in the empirical octane number models and the lack of ignition delay time data for gasolines reduces the surrogate formulation for SI engine autoignition to guess work.

Autoignition predictions of the three RON 90 fuels have been made for the LUPOE2-D conditions of Figure 3 using the three reduced mechanisms and the comparison to the observed knock onsets is presented in Figure 8. The three fuels are subject to that same end gas $p - T$ and equivalence ratio history. Since LUPOE2-D was operated in skip firing mode, all trapped residual gases were expunged, hence an ideal scavenging was assumed in the modelling. Owing to this, the differences in the predicted autoignition times are caused solely by differences in the autoignition chemistry of the three fuels. Across the range of studied conditions, the Andrae's model appears to perform consistently better than competitors, however more accurate PRF submodel in the Reitz model produces accurate autoignition prediction in the case of PRF90. Golovitchev's model predicts shorter delays in contrast to what can be seen in the ignition delay time calculations at the constant pressure, see Figure 1.

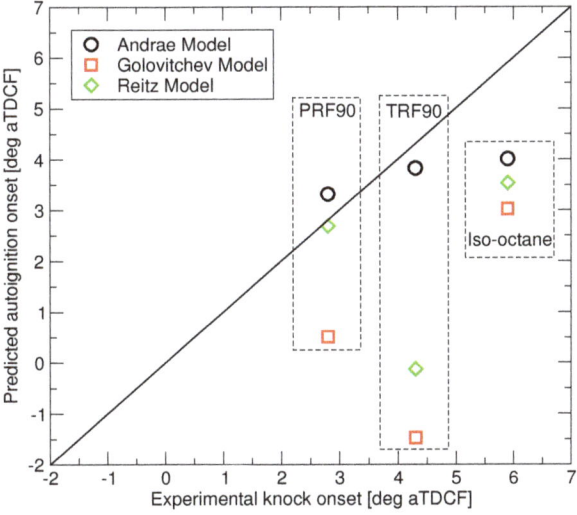

Figure 8. Predicted vs. measured autoignition onset for the three surrogate fuels.

Table 2 lists the possible surrogates for ULG90. The constraints discussed in Section 4 have been optimised for ULG90 to formulate a properties-based surrogate. Only the Reitz model has pathways for a naphthene, here cyclohexane, a suitable surrogate component. A composition-based surrogate is also studied whose composition is the same as that of the major gasoline constituent families. A tri-component TRF with the same RON and MON as ULG90 has also been studied. The autoignition onsets of the composition and properties based surrogates for the ULG90 $p - T$ history in LUPOE2-D have been simulated using the Reitz model. The autoignition onset of the TRF has been predicted by the three mechanisms. It is found that it is the TRF with Andrae model which predicts the closest autoignition to that of the ULG90. Among the two 4-component surrogates, it is the composition based surrogate which yields predictions closer to the observed knock onset.

Table 2. Key properties and volumetric composition of ULG90 and its surrogates. Predicted surrogate autoignition onsets for the ULG90 $p - T$ condition of Figure 3 have also been tabulated.

Gasoline Components	ULG90 % Vol.	Surrogate Components	Comp. Based	Props. Based	TRF
Iso-paraffins	60.0	Iso-octane	60.0	54.51	54.81
N-paraffins	8.0	N-heptane	8.0	16.35	16.74
Aromatics	21.0	Toluene	21.0	27.89	28.45
Naphthenes	11.0	Cyclohexane	11.0	1.26	-
RON	90		94.38	90	90
MON	84.7		89.65	84.7	84.67
A/F$_s$	14.69		14.56	14.69	14.49
Knock (aTDCF)	4.2°	AI * (Reitz)	2.87°	1.52°	1.16°
		AI (Golov.)	-	-	−1.06°
		AI (Andrae)	-	-	3.2°

* AI: Autoignition onset.

7. Conclusions

It has been demonstrated that reduced chemical kinetics can feasibly be used in the prediction of gasoline autoignition in a SI engine by approximating its properties by a surrogate. The choice of the mechanism and the surrogate affect greatly the prediction accuracy. The choice of the surrogate components is however limited by the availability of chemical kinetics schemes for their oxidation. Although the prediction of octane numbers is important in the formulation of a surrogate but matching of the RON and MON does not guarantee the correct reproduction of gasoline ignition behaviour in the surrogate as the modern SI engines operate at conditions much different form the RON and MON tests. Moreover, a compositional fidelity of the surrogate to that of the gasoline is also important in producing the correct autoignition behaviour in the surrogate.

The mechanisms generally over-predict the ignition delay time at low temperatures and pressure of about 40 bar. However, earlier autoignition onsets in an engine have been predicted by the three models; in particular the Golovitchev and Reitz models, indicating shorter ignition delay time predictions at lower pressures. There is still a need of validating reduced mechanisms at $p - T$ conditions which are found in modern downsized, turbocharged SI engines.

Author Contributions: Conceptualization and methodology: A.A.B.; software: A.F.K. and A.A.B.; validation, formal analysis, and investigation: A.F.K. and P.J.R.; data curation: A.A.B.; writing—original draft preparation A.F.K.; writing—review and editing: A.A.B.; visualization A.F.K.; supervision: A.A.B.

Acknowledgments: Authors gratefully acknowledge the financial support from EPSRC (Engineering and Physical Sciences Research Council), and Mahle Powertrain UK Ltd. Northampton UK, which funded a part of this work. Discussions with Jens Neumeister and Dave OudeNijeweme on the knocking tendency of modern high performance SI engines were useful. Many stimulating discussions with C.G.W. Sheppard, his inspiration and guidance during the experimental work were invaluable.

Conflicts of Interest: The authors declare no conflict of interest. The funders had no role in the design of the study; in the collection, analyses, or interpretation of data; in the writing of the manuscript, or in the decision to publish the results.

References

1. Khan, A.F.; Burluka, A.; Neumeister, J.; OudeNijeweme, D.; Freeland, P.; Mitcalf, J. Combustion and Autoignition Modelling in a Turbocharged SI Engine. *SAE Int. J. Engines* **2016**, *9*, 2079–2090. [CrossRef]
2. Bozza, F.; Siano, D.; Torella, E. Cycle-by-Cycle Analysis, Knock Modeling and Spark-Advance Setting of a "Downsized" Spark-Ignition Turbocharged Engine. *SAE Int. J. Engines* **2009**, *2*, 381–389. [CrossRef]
3. Mehl, M.; Faravelli, T.; Ranzi, E.; Giavazzi, F.; Scorletti, P.; Terna, D.; D'Errico, G.; Lucchini, T.; Onorati, A. *Kinetic Modelling Study of Octane Number and Sensitivity of Hydrocarbon Mixtures in CFR Engines*; SAE Technical Paper 2005-24-077; Consiglio Nazionale delle Ricerche: Rome, Italy, 2005. [CrossRef]
4. Andrae, J.; Head, R. HCCI experiments with gasoline surrogate fuels modeled by a semidetailed chemical kinetic model. *Combust. Flame* **2009**, *156*, 842–851. [CrossRef]
5. Huang, C.; Golovitchev, V.; Lipatnikov, A. Chemical Model of Gasoline-Ethanol Blends for Internal Combustion Engine Applications. In Proceedings of the SAE 2010 World Congress & Exhibition, Detroit, MI, USA, 13–15 April 2010; SAE Technical Paper 2010-01-0543; 2010.
6. Ra, Y.; Reitz, R.D. A combustion model for IC engine combustion simulations with multi-component fuels. *Combust. Flame* **2011**, *158*, 69–90. [CrossRef]
7. Knop, V.; Pera, C.; Duffour, F. Validation of a ternary gasoline surrogate in a CAI engine. *Combust. Flame* **2013**, *160*, 2067–2082. [CrossRef]
8. Vanhove, G.; Petit, G.; Minetti, R. Experimental study of the kinetic interactions in the low-temperature autoignition of hydrocarbon binary mixtures and a surrogate fuel. *Combust. Flame* **2006**, *145*, 521–532. [CrossRef]
9. Pera, C.; Knop, V. Methodology to define gasoline surrogates dedicated to auto-ignition in engines. *Fuel* **2012**, *96*, 59–69. [CrossRef]
10. Gauthier, B.; Davidson, D.; Hanson, R. Shock tube determination of ignition delay times in full-blend and surrogate fuel mixtures. *Combust. Flame* **2004**, *139*, 300–311. [CrossRef]
11. Knop, V.; Loos, M.; Pera, C.; Jeuland, N. A linear-by-mole blending rule for octane numbers of n-heptane/iso-octane/toluene mixtures. *Fuel* **2014**, *115*, 666–673. [CrossRef]
12. Mehl, M.; Chen, J.Y.; Pitz, W.J.; Sarathy, S.M.; Westbrook, C.K. An Approach for Formulating Surrogates for Gasoline with Application toward a Reduced Surrogate Mechanism for CFD Engine Modeling. *Energy Fuels* **2011**, *25*, 5215–5223. [CrossRef]
13. Kukkadapu, G.; Kumar, K.; Sung, C.J.; Mehl, M.; Pitz, W.J. Autoignition of gasoline and its surrogates in a rapid compression machine. *Proc. Combust. Inst.* **2013**, *34*, 345–352. [CrossRef]
14. Fikri, M.; Herzler, J.; Starke, R.; Schulz, C.; Roth, P.; Kalghatgi, G. Autoignition of gasoline surrogates mixtures at intermediate temperatures and high pressures. *Combust. Flame* **2008**, *152*, 276–281. [CrossRef]
15. Cancino, L.; Fikri, M.; Oliveira, A.; Schulz, C. Autoignition of gasoline surrogate mixtures at intermediate temperatures and high pressures: Experimental and numerical approaches. *Proc. Combust. Inst.* **2009**, *32*, 501–508. [CrossRef]
16. Khan, A.F.; Burluka, A.A. An investigation of various chemical kinetic models for the prediction of autoignition in HCCI engine. In Proceedings of the ASME 2012 Internal Combustion Engine Division Fall Technical Conference, Vancouver, BC, Canada, 23–26 September 2012; pp. 737–745.
17. Fieweger, K.; Blumenthal, R.; Adomeit, G. Self-ignition of S.I. engine model fuels: A shock tube investigation at high pressure. *Combust. Flame* **1997**, *109*, 599–619. [CrossRef]
18. Hartmann, M.; Gushterova, I.; Fikri, M.; Schulz, C.; Schießl, R.; Maas, U. Auto-ignition of toluene-doped n-heptane and iso-octane/air mixtures: High-pressure shock-tube experiments and kinetics modeling. *Combust. Flame* **2011**, *158*, 172–178. [CrossRef]
19. Davidson, D.; Gauthier, B.; Hanson, R. Shock tube ignition measurements of iso-octane/air and toluene/air at high pressures. *Proc. Combust. Inst.* **2005**, *30*, 1175–1182. [CrossRef]
20. Cash, J.R. Review Paper: Efficient numerical methods for the solution of stiff initial-value problems and differential algebraic equations. *Proc. R. Soc. Lond. Ser. A Math. Phys. Eng. Sci.* **2003**, *459*, 797–815. [CrossRef]
21. Burluka, A. Combustion in a Spark Ignition Engine. In *Handbook of Combustion, Voluome 3: Gaseous and Liquid Fuels*; Wiley-VCH Verlag: Weinheim, Germany, 2010; Chapter 13, pp. 389–414.
22. Ghosh, P.; Hickey, K.; Jaffe, S. Development of a detailed gasoline composition-based octane model. *Ind. Eng. Chem. Res.* **2006**, *45*, 337–345. [CrossRef]

23. Foong, T.M.; Morganti, K.J.; Brear, M.J.; da Silva, G.; Yang, Y.; Dryer, F.L. The octane numbers of ethanol blended with gasoline and its surrogates. *Fuel* **2014**, *115*, 727–739. [CrossRef]
24. Morgan, N.; Smallbone, A.; Bhave, A.; Kraft, M.; Cracknell, R.; Kalghatgi, G. Mapping surrogate gasoline compositions into RON/MON space. *Combust. Flame* **2010**, *157*, 1122–1131. [CrossRef]
25. Roberts, P.; Sheppard, C. The Influence of Residual Gas NO Content on Knock Onset of Iso-Octane, PRF, TRF and ULG Mixtures in SI Engines. *Sae Int. J. Engines* **2013**, *6*, 2028–2043. [CrossRef]
26. Liu, K.; Burluka, A.; Sheppard, C. Turbulent flame and mass burning rate in a spark ignition engine. *Fuel* **2013**, *107*, 202–208. [CrossRef]

© 2019 by the authors. Licensee MDPI, Basel, Switzerland. This article is an open access article distributed under the terms and conditions of the Creative Commons Attribution (CC BY) license (http://creativecommons.org/licenses/by/4.0/).

Article

Numerical Investigation of Pressure Influence on the Confined Turbulent Boundary Layer Flashback Process

Aaron Endres * and Thomas Sattelmayer

Lehrstuhl für Thermodynamik, Technische Universität München, 85747 Garching, Germany
* Correspondence: endres@td.mw.tum.de

Received: 14 June 2019; Accepted: 28 July 2019; Published: 1 August 2019

Abstract: Boundary layer flashback from the combustion chamber into the premixing section is a threat associated with the premixed combustion of hydrogen-containing fuels in gas turbines. In this study, the effect of pressure on the confined flashback behaviour of hydrogen-air flames was investigated numerically. This was done by means of large eddy simulations with finite rate chemistry as well as detailed chemical kinetics and diffusion models at pressures between 0.5 bar and 3 bar. It was found that the flashback propensity increases with increasing pressure. The separation zone size and the turbulent flame speed at flashback conditions decrease with increasing pressure, which decreases flashback propensity. At the same time the quenching distance decreases with increasing pressure, which increases flashback propensity. It is not possible to predict the occurrence of boundary layer flashback based on the turbulent flame speed or the ratio of separation zone size to quenching distance alone. Instead the interaction of all effects has to be accounted for when modelling boundary layer flashback. It was further found that the pressure rise ahead of the flame cannot be approximated by one-dimensional analyses and that the assumptions of the boundary layer theory are not satisfied during confined boundary layer flashback.

Keywords: large eddy simulation; confined; boundary layer flashback; turbulent combustion; hydrogen

---

1. Introduction

The combustion of hydrogen-rich fuels instead of pure hydrocarbon fuels in gas turbines is one possible measure to reduce anthropogenic CO_2 emissions in power production. Hydrogen-rich synthetic fuels can be produced by reforming or gasification of hydrocarbon fuels. The low carbon content of these synthetic fuels results in low CO_2 production during the combustion process. The CO_2 emissions that accrue during the synthesis process on the other hand can be captured and stored prior to the combustion process [1,2]. Another measure to decrease CO2 emission in power production is the exploitation of renewable energy resources. Power production from renewable resources and power demand generally show strong fluctuations. Hydrogen can be produced from the electrolysis of water in times of excess power output. By using the produced hydrogen as a fuel in gas turbines, the imbalance of power production and demand can be compensated without additional CO_2 emission [1,2].

In order to meet the strict regulations of NOx emissions, lean premixed combustion is usually applied in gas turbines [3]. By mixing fuel and oxidizer upstream of the combustion chamber, combustible fuel-air mixtures exist upstream of the combustion chamber. This allows for flame flashback into the premixing channel of the burner, which potentially causes critical damage to the premixing channel and requires engine shutdown. Boundary layer flashback (BLF) is upstream flame propagation which takes place in the low velocity region of the boundary layer. Hydrogen flames show high burning velocities and a high reactivity near walls. This makes hydrogen combustion in gas turbines prone to BLF [4].

Due to the risk associated with BLF, it is essential to be aware of the flashback limits and their influencing parameters. The influence of preheat temperature [5,6], flame confinement [7] and acoustic oscillations [8] on the flashback behaviour, for example, have been investigated experimentally. Furthermore, Daniele et al. [5] and Kalantari et al. [6] investigated the influence of pressure on the flashback limits of unconfined turbulent lean premixed flames with high hydrogen content. They observed that the flashback equivalence ratio Φ_{FB} decreases with pressure p and follows the power law $\Phi_{FB} \propto p^{-n}$. Depending on the operating conditions in the experiments, the pressure exponent n lay between 0.27 and 0.51. They concluded that this strong increase in flashback tendency with increasing pressure is mainly caused by the decrease in quenching distance. The change in turbulent flame speed with pressure on the contrary only had a minor effect on the flashback limits.

Hoferichter and Sattelmayer [8] showed that the flashback limits of unconfined flames, which are acoustically excited, approach the flashback limits of confined flames, which are not acoustically excited. The confined case of BLF is thus the limiting of both cases regarding the flashback safety of burners. Although it is of high technical relevance, the pressure influence on the flashback process of confined turbulent flames has not been investigated yet. Furthermore, only the global influence of pressure on unconfined flashback limits has been investigated so far. The influence of pressure on parameters, such as the quenching distance, had to be estimated from analytical considerations [5,6].

Hoferichter et al. [9] developed an analytical model to predict atmospheric confined boundary layer flashback limits. In their model, flashback is assumed to occur when the average flame front causes a pressure rise which is large enough to cause boundary layer separation. The adapted Stratford separation criterion [10],

$$\frac{\Delta p_{max}}{\rho_u U_{FB}^2} = 0.0975^{\frac{2}{3}}, \tag{1}$$

is applied for the prediction of boundary layer separation and the confined flashback limits. U_{FB} is the channel centerline velocity at which boundary layer separation and flashback are expected. The pressure rise Δp_{max} ahead of the flame front is estimated from one-dimensional considerations. The conservation of mass and momentum and assuming equal pressures far upstream and downstream of the flame results in the pressure rise

$$\Delta p_{max} = \rho_u S_{t,max}^2 \left(\frac{T_{ad}}{T_u} - 1\right), \tag{2}$$

which is dependent on the fresh gas density ρ_u, the maximum turbulent flame speed $S_{t,max}$ and the ratio of unburnt gas temperature T_u to adiabatic flame temperature T_{ad}. The turbulent flame speed is estimated from the laminar flame speed, flame stretch and the maximum turbulent fluctuation velocity [9]. The pressure rise from Equation (2) and the Stratford separation criterion (1) lead to the flashback criterion

$$\frac{S_{t,max}^2}{U_{FB}^2} \left(\frac{T_{ad}}{T_u} - 1\right) = 0.0975^{\frac{2}{3}}, \tag{3}$$

or

$$U_{FB} = 0.0975^{-\frac{1}{3}} S_{t,max} \sqrt{\frac{T_{ad}}{T_u} - 1}. \tag{4}$$

This criterion implies that the flashback centerline velocity is only dependent on the maximum turbulent flame speed and the gas expansion ratio given by the temperatures T_{ad} and T_u. This contradicts the observation of Daniele et al. [5] and Kalantari et al. [6], who found that the quenching distance has a strong influence on unconfined BLF limits. Furthermore, the one-dimensional approximation of the pressure rise ahead of the flame front contradicts the findings of Eichler [11] from two-dimensional simulations of confined laminar BLF. He observed that the velocity field and the pressure rise ahead of the flame are of two-dimensional nature, which prohibits the prediction of the pressure rise based on one-dimensional considerations.

Several numerical simulations of turbulent flashback of hydrogen-rich fuels have been presented in recent years [4,12–14]. These simulations significantly improved the understanding of the BLF process. Endres and Sattelmayer [15] recently presented first simulations that accurately reproduced confined flashback limits of turbulent hydrogen-air flames at atmospheric pressure. They applied large eddy simulations (LES) with finite rate chemistry and a detailed species diffusion model. The study also showed that under atmospheric conditions, upstream flame propagation occurs when the flow separation zone size ahead of the propagating flame front is significantly larger than the quenching distance of the flame.

A similar LES model as previously presented for atmospheric calculations [15] is applied in the present work. This approach allows for a detailed analysis of the flashback process at different pressure levels. The comparison of the numerically obtained flashback limits at atmospheric pressure with experimental flashback limits [16] ensures that the chosen simulation approach is capable of accurately reproducing the flashback process. Flashback can then be investigated at pressures between 0.5 bar and 3 bar, where no experimental data is available. Although changing the pressure leads to altered molecular viscosities and Reynolds numbers, the atmospheric channel flow approaching the flame is left unchanged in this work. Comparing the results at different pressure levels allows for the investigation of the influence of average quantities, such as the maximum turbulent flame speed and the average flow deflection, on the flashback limits. This is followed by an analysis of the influence of local quantities, such as the local quenching distance and the local separation zone size, on the flashback process. Finally, the results are used to investigate the applicability of the one-dimensional pressure estimation (2) and the Stratford separation criterion (1) for BLF prediction.

2. Numerical Model

Despite introducing in-situ adaptive tabulation and a low-Mach number approximation in the reactive solver, the numerical model is similar to the numerical model of Endres and Sattelmayer [15]. For the sake of completeness, a shortened presentation of the numerical model will be given here. For more details on the model, the reader is referred to the article of Endres and Sattelmayer [15].

The numerical model for inert and reactive simulations is implemented in the framework of the open source computational fluid dynamics package OpenFOAM [17]. The governing equations are solved with a finite volume discretization. Derivatives in time are discretised by a second order accurate quadratic backward differencing scheme. Spatial discretization is handled by the second order accurate central differencing scheme and by a stabilised central differencing scheme [18].

2.1. Inert Simulations

An accurate inert base flow is a basic requirement for the atmospheric flashback simulations in order to precisely predict the experimental BLF limits. In the experiments, the velocity profile at the beginning of the flame stabilization section well reproduced the characteristics of a fully developed channel flow [16]. Therefore, the inert simulations also aim at accurately predicting the mean velocity profile and the velocity fluctuations of a fully developed channel flow.

The incompressible OpenFOAM solver pimpleFoam is used for the inert simulations. In this solver, the incompressible Navier-Stokes equations are solved with a segregated approach. The time step size is automatically adapted in order to keep the local non-acoustic Courant number below 0.8. The OpenFOAM version of the Smagorinsky model [15,17,19,20] is used for modelling the subgrid scale (SGS) kinematic viscosity ν_{SGS}. The chosen model constants are equivalent to a Smagorinsky constant of $c_S = 0.168$ and are close to the original value of $c_S = 0.17$ proposed by Lilly [21]. The filter width Δ is approximated by the cube root of the local cell volume. Van Driest damping is applied to the filter width in order to account for low Reynolds effects near walls and to ensure a vanishing SGS viscosity at walls [22,23].

The numerical flashback results at atmospheric pressure are compared with experimental flashback results in a rectangular channel [7]. Bulk velocities U_b of $10\,\mathrm{m\,s^{-1}}$, $20\,\mathrm{m\,s^{-1}}$ and $30\,\mathrm{m\,s^{-1}}$ are investigated. For the representation of the experimental rectangular channel geometry, the computational domain for the inert simulations is a channel with a half height of $\delta = 8.75\,\mathrm{mm}$, a length of 6δ and a width of 3δ. Cyclic boundary conditions are prescribed at the front and back patches in order to obtain a quasi two-dimensional configuration. Cyclic boundary conditions are also set at the inlet and outlet patches. This allows for full development of the channel flow. In order to obtain the desired bulk velocities, an average streamwise pressure gradient is automatically prescribed along the channel. The walls are modelled as no-slip walls without applying a wall model. Therefore, small wall-normal cell sizes $y_1^+ < 0.8$ of the cells adjacent to walls are chosen. The wall-normal cell size is gradually increased towards the channel centerline where the cell size is ten times the cell size at the wall. The resulting maximum cell sizes and cell counts in streamwise (x), wall-normal and spanwise (z) direction are given in Table 1. All cell sizes are made non-dimensional by the kinematic viscosity ν and the theoretical friction velocity calculated with the empirical correlation developed by Pope [24],

$$u_{\tau,Pope} = 0.09 \frac{y}{\delta} \mathrm{Re}^{0.88}. \tag{5}$$

The kinematic viscosity was fixed to $\nu = 1.8 \times 10^{-5}\,\mathrm{m^2\,s^{-1}}$.

This numerical setup resulted in accurate average velocity profiles and turbulent kinetic energy profiles compared to literature DNS results [15]. It was furthermore shown by Endres and Sattelmayer [15] that the flashback limits obtained from the atmospheric reactive simulations were insensitive to changes in the domain size of the inert simulations and to changes in the resolution of the inert base flow. The time-dependent velocity profiles from the inert simulations were therefore again sampled and projected onto the inlet of the domain of the reactive simulations.

Table 1. Simulation parameters for inert large eddy simulations (LES).

U_b	Re	$u_{\tau,Pope}$	$\mathrm{Re}_{\tau,Pope}$	y_1^+	Δx^+	Δy^+	Δz^+	$N_x \times N_y \times N_z$
$10\,\mathrm{m\,s^{-1}}$	9722	0.598	290.7	0.63	13.42	12.52	8.90	$130 \times 118 \times 98$
$20\,\mathrm{m\,s^{-1}}$	19,444	1.101	535.2	0.66	13.72	13.11	9.02	$234 \times 208 \times 178$
$30\,\mathrm{m\,s^{-1}}$	29,167	1.573	764.4	0.79	18.35	16.24	12.07	$250 \times 240 \times 190$

The chosen viscosity well represents the viscosity of the investigated hydrogen-air mixtures at atmospheric pressure. By changing the domain pressure to subatmospheric or elevated pressure, the kinematic viscosity of the hydrogen-air mixtures is changed. This modifies the flow Reynolds number and the channel flow characteristics. In the present work, the inert viscosity and the inert base flow are however left unchanged compared to the atmospheric calculations. The effect of pressure on the inert velocity profile is thereby eliminated and the sole effect of pressure on the combustion and flow separation properties is investigated.

2.2. Reactive Simulations

The numerical model for the reactive simulations applied by Endres and Sattelmayer [15] was proven to accurately predict the flashback behaviour of atmospheric turbulent hydrogen-air flames. However, it was later noted that despite applying a wave-transmissive boundary condition, acoustic pressure oscillations occured in the domain, which possibly influenced the results. A low Mach number approximation

$$\rho = \frac{p_0}{R_s T} \tag{6}$$

is thus applied to the compressible governing equations [25]. By calculating the gas density ρ from a constant thermodynamic pressure p_0, the specific gas constant R_s and temperature T, density changes are decoupled from pressure changes and acoustic oscillations are prevented. The constant thermodynamic pressure is also used to calculate the molecular gas properties as well as the chemical reaction rates. The Navier-Stokes equations on the other hand are solved for the dynamic pressure p_d. In addition, the transport equation for sensible enthalpy and the Favre-filtered transport equations

$$\frac{\partial \left(\bar{\rho}\widetilde{Y}_k\right)}{\partial t} + \frac{\partial \left(\bar{\rho}\widetilde{u}_i\widetilde{Y}_k\right)}{\partial x_i} + \frac{\partial \overline{j_{k,i}}}{\partial x_i} - \frac{\partial}{\partial x_i}\left(\bar{\rho}\frac{\nu_{SGS}}{Sc_{SGS}}\frac{\partial \widetilde{Y}_k}{\partial x_i}\right) = \overline{\dot{\omega}_k} \tag{7}$$

for the mass fractions Y_k of each species k are solved in order to include chemical reactions. The SGS Schmidt number Sc_{SGS} is assumed to be unity. The Smagorinsky model is used for modelling the SGS viscosity ν_{SGS}. The chemical reaction rate $\dot{\omega}_k$ is obtained from a system of chemical reactions of Arrhenius type solved by the Seulex ordinary differential equations (ODE) solver [26]. The detailed chemical kinetics mechanism of Burke et al. [27] is used for modelling the chemistry of hydrogen combustion. This mechanism consists of 9 species and 23 reactions. In order to increase the speed of the calculations compared to the calculations in Reference [15], in-situ adaptive tabulation [28,29] is applied in the present work. By storing the results of the direct solution of the chemical reaction mechanism, a look-up table is created for mapping the gas composition, pressure and temperature from the previous solver iteration to the current iteration. The number of computationally costly ODE integrations is thereby kept at a minimum. The reaction mapping is approximated linearly by a mapping gradient matrix. The maximum error of this linear approximation is set to 10^{-4}.

The filtered chemical reaction rate $\overline{\dot{\omega}_k}$ of species k is assumed to be equivalent to the reaction rate directly obtained from the filtered temperature, pressure and species mass fraction distribution. This approach is usually referred to as implicit LES (ILES) [25,30] or quasi-laminar combustion model [31]. The subgrid turbulence-chemistry interaction is neglected in ILES. Therefore, it has to be ensured that the flame either has a laminar character at the subgrid scales [15,32] or that the LES cells are perfectly stirred by turbulent fluctuations [25].

Endres and Sattelmayer [15] showed that a detailed diffusion model is essential for the prediction of correct flame shapes and flashback limits. The diffusion mass flux $j_{k,i}$ is therefore calculated as the sum of the diffusion mass flux due to species mass fraction gradients $j^Y_{k,i}$ and the Soret diffusion mass flux $j^T_{k,i}$ due to temperature gradients.

$$j_{k,i} = j^Y_{k,i} + j^T_{k,i} = -\rho D_k \frac{\partial Y_k}{\partial x_i} - D^T_k \frac{1}{T}\frac{\partial T}{\partial x_i} \tag{8}$$

The mixture diffusion coefficient D_k of species k is obtained from the binary diffusion coefficients $D_{j,k}$ of species j through species k and a gas composition dependent mixture equation [33,34]. The binary diffusion coefficients are in turn obtained from an empricial equation based on the kinetic gas theory [35,36]. The Soret diffusion coefficients D^T_k are calculated according to an empirical correlation [37]. The mixture thermal diffusivity α for the transport equation of sensible enthalpy is calculated from the thermal diffusivities of each species by applying a gas composition dependent mixture equation [34,38]. The species thermal diffusivities are approximated by a third degree logarithmic polynomial [15]. The polynomial coefficients are calculated beforehand by applying a least squares fit to the species thermal diffusivities obtained with Cantera [38]. The kinematic viscosity ν is obtained from the Sutherland law [39] with the Sutherland parameter $A_S = 1.67121\,\mathrm{kg\,m^{-1}\,s^{-2}\,K^{-0.5}}$ and the Sutherland temperature $T_S = 170.672\,\mathrm{K}$.

Hoferichter and Sattelmayer [8] showed the flashback limits of confined BLF are of high technical relevance for unconfined flames with acoustic excitation. The present simulations therefore aim at reproducing the confined flashback limits presented by Eichler et al. [16]. As in the atmospheric calculations of Endres and Sattelmayer [15], the computational domain depicted in Figure 1 represents a rectangular channel with a length of 6δ and a width of 3δ for $U_b = 10\,\mathrm{m\,s^{-1}}$ and a width of 2.23δ for $U_b = 20\,\mathrm{m\,s^{-1}}$ and $U_b = 30\,\mathrm{m\,s^{-1}}$. The time dependent velocity profile is prescribed at the inlet and a constant dynamic pressure is prescribed at the outlet. In order to obtain the flow characteristics of an infinitely wide channel, cyclic boundary conditions are prescribed at the front and back patch. The lower wall is modelled as an isothermal wall at 293.15 K up to $x \leq 3\delta$ and as an adiabatic wall for $x > 3\delta$, in order to represent the cooled steel walls followed by a ceramic tile for flame stabilization in the experiments [16]. The base mesh from the inert calculations is left unchanged. In order to well resolve the stable and propagating flame front, the mesh is automatically refined by adaptive mesh refinement. The hexahedral cells are split in each spatial direction resulting in eight smaller hexahedral cells. This cell split is applied to a cell if the change $\Delta_{H_2} = \left| \nabla Y_{H_2} V_c^{1/3} \right|$ of hydrogen mass fraction in one cell of volume V_c is larger than 10 % of the unburnt hydrogen mass fraction. The maximum number of splits per cell is limited to one or two splits in the investigated cases.

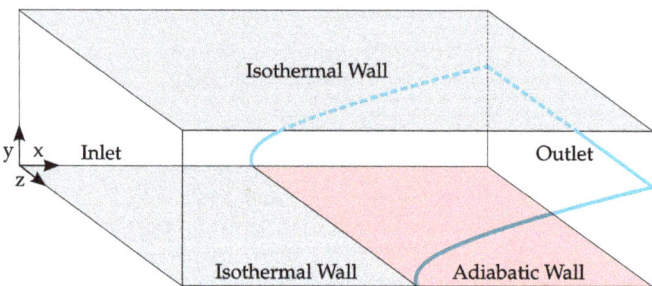

Figure 1. Computational domain and stable flame shape for reactive LES.

The approach for obtaining flashback limits is similar to the atmospheric calculations in Reference [15]. A constant velocity is assumed in the channel at the beginning of the calculations. The species mass fractions of hydrogen, oxygen and nitrogen are set to an equivalence ratio at which no flashback is expected. The flame front is initialized by setting the temperature and species mass fractions in a rectangular box with a height of 1 mm to the burnt gas values. As soon as the velocity fluctuations prescribed at the inlet reach the flame front, the flame front is wrinkled and stabilizes at the leading edge of the adiabatic wall. When the flame wrinkling is fully developed, the equivalence ratio Φ at the inlet is increased by a small step of 0.02 to 0.05. This is repeated until the flame front propagates upstream along the isothermal wall without being completely washed out. This equivalence ratio is defined as the flashback equivalence ratio Φ_{FB}.

2.3. Les Regimes for Hydrogen Combustion

The requirement for the applicability of the ILES approach is that the SGS flame either has a laminar structure, or that the reaction and mixing processes are not influenced by SGS fluctuations. The LES combustion regimes developed by Pitsch and Duchamp de Lageneste [32] can be used to point out the laminar character of the SGS flame structure in the presented simulations. In the derivation of the original LES combustion regime diagram, Pitsch and Duchamp de Lageneste assume that the diffusivity D of the gas mixture equals the kinematic viscosity ν. This assumption is however invalid for hydrogen combustion. In the following, the LES combustion regimes are therefore analyzed for the general case, where $D \neq \nu$.

The combustion regimes are defined by the Karlovitz number Ka and the ratio of the LES filter width Δ to the laminar flame length scale $l_f = D/s_{l,0}$ [40], where $s_{l,0}$ is the laminar flame speed. The Karlovitz number is defined as the ratio of the flame time scale $t_f = l_f^2/D$ to the time scale $t_\eta = \eta^2/\nu$ of the Kolmogorov eddies. The Kolmogorov length scale $\eta = \left(\nu^3/\epsilon \right)^{0.25}$ can be approximated by the kinematic viscosity ν and the dissipation rate $\epsilon = c_\epsilon k_{SGS}^{1.5} \Delta^{-1}$. Here, k_{SGS} is the SGS turbulent kinetic energy and c_ϵ is a model constant. This results in the diffusivity and viscosity dependent Karlovitz number calculated from LES quantities,

$$Ka = \sqrt{\frac{c_\epsilon k_{SGS}^{1.5} D^2}{\Delta s_{l,0}^4 \nu}}. \tag{9}$$

The wrinkled flamelet regime and the corrugated flamelet regime are separated by the Gibson length scale. The Gibson length scale is defined by the equality of the laminar flame speed and the SGS velocity fluctuation $\sqrt{k_{SGS}}$. In the wrinkled flamelet regime, the SGS velocity fluctuation is smaller or equal to the laminar flame speed. According to Peters [40], fluctuations with a velocity smaller than the laminar flame speed do not wrinkle the flame front. The SGS flame therefore has a laminar character in the wrinkled flamelet regime while it has a non-laminar character in the corrugated flamelet regime. With the assumption that $D = \nu$, the condition $s_{l,0} = \sqrt{k_{SGS}}$ for the Gibson length scale leads to a mixture independent relation between the Karlovitz number and Δ/l_f. For hydrogen however, $D \neq \nu$ and the condition for the Gibson length scale lead to

$$\frac{\Delta}{l_f} = Ka^{-2} \frac{D}{\nu}, \tag{10}$$

when assuming $c_\epsilon \approx 1$. This takes the viscosity and diffusivity of the gas mixture into account when evaluating the SGS combustion regime. This results in the different lines between the wrinkled flamelet and corrugated flamelet regime in Figures 2 and 3.

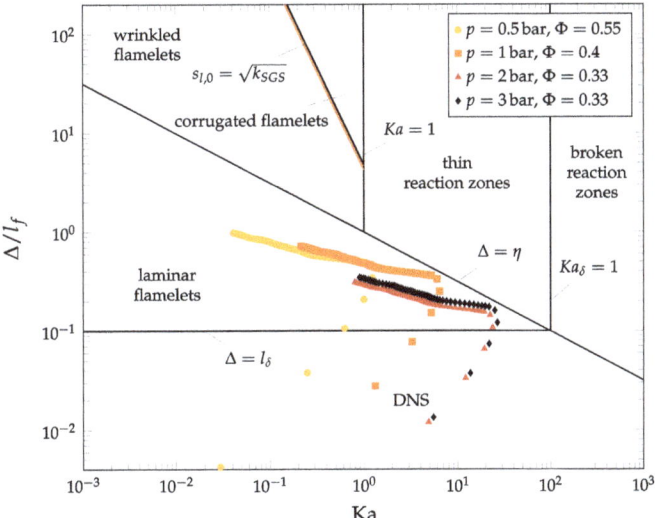

Figure 2. LES combustion regimes for $U_b = 10\,\mathrm{m\,s^{-1}}$.

Figures 2 and 3 show the SGS combustion regimes at different pressure levels for the highest corresponding equivalence ratios investigated in the current work. Each data point represents one cell along the y-axis of the unrefined mesh. The velocity fluctuations are obtained from the inert calculations and the laminar flame speed is calculated from the correlation developed by Böck [41] for atmospheric flames. The influence of pressure on the laminar flame speed is taken into account by the proportionality $s_{l,0} \propto p^m$. The pressure exponent m is obtained from the results of Bradley et al. [42]. Following Bechtold and Matalon [43], the Lewis number Le of gas mixtures can be obtained from the species diffusivities and a mixture and flame temperature dependent weight function.

$$Le = \frac{\alpha}{D} = 1 + \frac{\left(\frac{\alpha_{O_2}}{D_{O_2}} - 1\right) + \left(\frac{\alpha_{H_2}}{D_{H_2}} - 1\right)\left(1 + Ze\left(\frac{1}{\Phi} - 1\right)\right)}{2 + Ze\left(\frac{1}{\Phi} - 1\right)} \tag{11}$$

For this calculation, the species diffusivities $D_{H_2} = 7.29 \times 10^{-5}\,\text{m}^2\,\text{s}^{-1}$ and $D_{O_2} = 1.88 \times 10^{-5}\,\text{m}^2\,\text{s}^{-1}$ and species thermal diffusivities $\alpha_{H_2} = \alpha_{O_2} = 2.25 \times 10^{-5}\,\text{m}^2\,\text{s}^{-1}$ are assumed to be mixture independent. The Zeldovich number

$$Ze = \frac{E(T_{ad} - T_u)}{RT_{ad}^2} \tag{12}$$

is calculated from the adiabatic flame temperature T_{ad}, the unburnt gas temperature $T_u = 293.15\,\text{K}$ the constant activation energy $E = 20\,\text{kcal mol}^{-1} = 8.368 \times 10^4\,\text{kJ kmol}^{-1}$ and the universal gas constant R. With the resulting Lewis numbers and the actual mixture thermal diffusivities α obtained from Cantera, the mixture diffusivities D can be computed using the Lewis number definition $Le = \alpha/D$. The input and resulting parameters for the diffusivity calculation for all cases are listed in Table A1 in Appendix A.

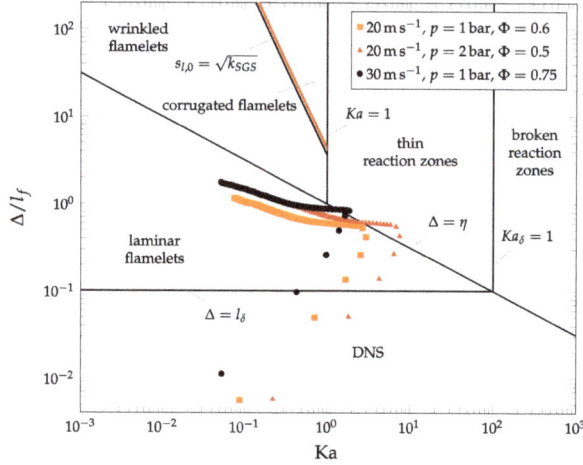

Figure 3. LES combustion regimes for $U_b = 20\,\text{m s}^{-1}$ and $U_b = 30\,\text{m s}^{-1}$.

It is evident from Figure 2 that all cases at $U_b = 10\,\text{m s}^{-1}$ lie within the laminar flamelet and the DNS regimes. In the laminar flamelet regime the filter width Δ is smaller than the Kolmogorov length scale η. The SGS flow is therefore approximately laminar and no SGS turbulence fluctuations will have an influence on the combustion process [32]. At $U_b = 20\,\text{m s}^{-1}$ and $p = 1\,\text{bar}$, all cells also lie within the laminar flamelet and DNS regimes. At $U_b = 20\,\text{m s}^{-1}$ and $p = 2\,\text{bar}$ as well as at $U_b = 30\,\text{m s}^{-1}$ and $p = 1\,\text{bar}$, some cells lie within the thin reaction zones regime. In this regime, the SGS turbulent eddies are small enough to penetrate the flame preheat zone and affect the mixing processes in the preheat zone [32]. The cells in the thin reaction zone regime should therefore not be modeled with the ILES combustion model. This has to be taken into account when analyzing the results of the two cases. Especially for $U_b = 30\,\text{m s}^{-1}$ and $p = 1\,\text{bar}$, the majority of the cells however lie within the laminar flamelet and DNS regimes and show a laminar SGS flame structure. In order to also investigate the pressure influence for different bulk velocities, the $U_b = 20\,\text{m s}^{-1}$ case at $p = 2\,\text{bar}$ and the $U_b = 30\,\text{m s}^{-1}$ case at $p = 1\,\text{bar}$ are included here nevertheless. For all other cases the ILES approach is valid and the results are expected to deliver accurate results.

3. Results

The presented approach leads to flashback limits at different pressure levels with the unaltered inert base flow. In the following, the pressure influence on the flashback limits is investigated. This is followed by an analysis of the macroscopic flame structure and the flame resolution resulting from the macroscopic flame structure. Finally, the influence of pressure on the combustion and flow separation characteristics is investigated on the basis of average pressure and velocity fields and on the basis of local flame quenching and boundary layer separation.

3.1. Pressure Influence on Confined Flashback Limits

At atmospheric pressure, confined boundary layer flashback is modeled at bulk velocitites of $10\,\mathrm{m\,s^{-1}}$, $20\,\mathrm{m\,s^{-1}}$ and $30\,\mathrm{m\,s^{-1}}$. It was shown by Endres and Sattelmayer [15] that one cell split is sufficient to resolve the flame front at $U_b = 10\,\mathrm{m\,s^{-1}}$ and $U_b = 20\,\mathrm{m\,s^{-1}}$. Applying two cell splits by the adaptive mesh refinement algorithm at $U_b = 20\,\mathrm{m\,s^{-1}}$ and $U_b = 30\,\mathrm{m\,s^{-1}}$ is furthermore prohibitive due to the associated computational costs. The adaptive mesh refinement is therefore set to a maximum of one cell split for the atmospheric simulations. With this refinement strategy, the flame is stable up to equivalence ratios of 0.38, 0.55 and 0.7 for bulk velocities of $10\,\mathrm{m\,s^{-1}}$, $20\,\mathrm{m\,s^{-1}}$ and $30\,\mathrm{m\,s^{-1}}$, respectively. Although flame tongues already form and propagate at these equivalence ratios, they are generally flushed out by high speed flow structures after a few millimeters of propagation. When further increasing the equivalence ratios to 0.4, 0.6 and 0.75 for bulk velocities of $10\,\mathrm{m\,s^{-1}}$, $20\,\mathrm{m\,s^{-1}}$ and $30\,\mathrm{m\,s^{-1}}$, flame propagation can no longer be prevented by the approaching flow and the flame is not fully flushed out of the isothermal channel section. It is noted that the flashback limit for $U_b = 20\,\mathrm{m\,s^{-1}}$ and atmospheric pressure has increased from $\Phi = 0.55$ to $\Phi = 0.6$ compared to the results presented in Reference [15]. This is due to the changed definition of the flashback limit. Flame propagation was previously declared flashback if the flame was capable of propagating for more than 5 mm [15]. For $U_b = 20\,\mathrm{m\,s^{-1}}$ and $\Phi = 0.55$ it is however observed that the flame may still be flushed out after a propagation of more than 5 mm. Flame propagation is therefore now only declared flashback if the flame is not fully flushed out of the isothermal channel section.

In Figure 4, the numerically obtained stable equivalence ratios and the equivalence ratios at flashback are compared with experimental flashback limits obtained by Eichler et al. [16] with a similar channel configuration. In the experiments, flashback was first observed approximately at $\Phi = 0.35$, $\Phi = 0.54$ and $\Phi = 0.71$. It is evident that despite a slight underestimation of the flashback tendency for bulk velocities of $10\,\mathrm{m\,s^{-1}}$ and $20\,\mathrm{m\,s^{-1}}$, the numerical simulations are capable of reproducing the experimental flashback limits with reasonable accuracy. The simulations are therefore capable of representing the flashback process not only qualitatively but also quantitatively. This is in accordance with the findings in Reference [15], where no low Mach number approximation was applied.

In addition to the atmospheric simulations, flashback simulations are performed at a subatmospheric pressure of 0.5 bar and at slightly elevated pressures of 2 bar and 3 bar. At $p = 0.5\,\mathrm{bar}$, bulk velocities of $10\,\mathrm{m\,s^{-1}}$ and $20\,\mathrm{m\,s^{-1}}$ are numerically investigated. As the flame thickness is expected to be larger at lower pressure levels, no additional mesh refinement is necessary compared to the atmospheric simulations. Only one cell split is thus applied to the base mesh by the adaptive mesh refinement algorithm. The flame thickness is however reduced at higher pressure levels. Two cell splits are therefore applied at $p = 2\,\mathrm{bar}$ and a bulk velocity of $10\,\mathrm{m\,s^{-1}}$. As already mentioned for the atmospheric simulations, two cell splits can not be applied to simulations with $U_b = 20\,\mathrm{m\,s^{-1}}$ due to limited computational resources. A maximum of only one cell split is therefore applied at $U_b = 20\,\mathrm{m\,s^{-1}}$ and $p = 2\,\mathrm{bar}$. As one cell split is not expected to be sufficient for $p = 3\,\mathrm{bar}$ and $U_b = 20\,\mathrm{m\,s^{-1}}$, the investigation of the highest pressure level is limited to $U_b = 10\,\mathrm{m\,s^{-1}}$. Here, the initial inert mesh is refined in x-direction by increasing the cell count in x-direction by a factor of 1.5. The adaptive mesh refinement additionally refines the mesh by applying a maximum of two cell splits.

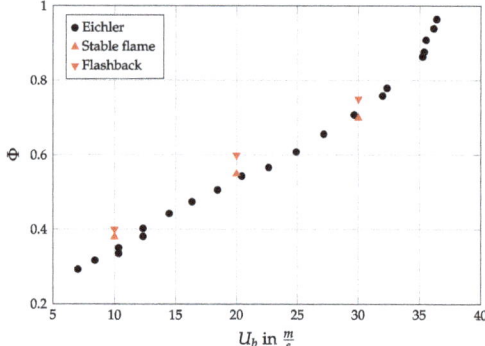

Figure 4. Atmospheric flashback limits obtained from the reactive LES compared to experimental flashback limits of Eichler et al. [16].

The flashback limits at different pressure levels resulting from this refinement strategy are depicted in Figure 5. Here, the flashback limit is defined as the first equivalence ratio, where the flame is not flushed out of the isothermal section of the channel. The error bars illustrate the fact that flashback might already occur at any equivalence ratio between the last stable simulation and the first simulation in which flashback occurs. For $p = 0.5\,\text{bar}$ and $U_b = 20\,\text{m}\,\text{s}^{-1}$, no flashback was observed up $\Phi = 1.7$. The laminar flame speed obtained from Cantera simulations reaches its maximum below this equivalence ratio. Increasing the equivalence ratio beyond $\Phi = 1.7$ does therefore not increase flashback propensity and flashback is not expected to occur for $p = 0.5\,\text{bar}$ and $U_b = 20\,\text{m}\,\text{s}^{-1}$. From Figure 5 it is evident that increasing the pressure leads to lower flashback limits. Only the case at $p = 3\,\text{bar}$ and $U_b = 10\,\text{m}\,\text{s}^{-1}$ shows an identical flashback limit compared to the lower pressure case at $p = 2\,\text{bar}$. At $U_b = 10\,\text{m}\,\text{s}^{-1}$, the pressure exponent of the dependency $\Phi_{FB} \propto p^{-n}$ is $n = 0.46$, $n = 0.37$, $n = 0.29$ at pressures of 1 bar, 2 bar and 3 bar compared to the flashback limit at $p = 0.5\,\text{bar}$. At $U_b = 20\,\text{m}\,\text{s}^{-1}$, the pressure exponent is $n = 0.26$ at $p = 2\,\text{bar}$ compared to the flashback limit at $p = 1\,\text{bar}$. In literature, pressure exponents of unconfined flashback have been found to lie between $n = 0.27$ and $n = 0.51$ for H_2-CO-air flames and between $n = 0.43$ and $n = 0.49$ for H_2-air flames [5,6]. The numerically obtained pressure exponents for confined flames are thus similar to experimentally obtained pressure exponents for unconfined flames. In the confined simulations, there is however no unique pressure exponent for one bulk velocity. Instead, the pressure exponent for $U_b = 10\,\text{m}\,\text{s}^{-1}$ decreases with increasing pressure.

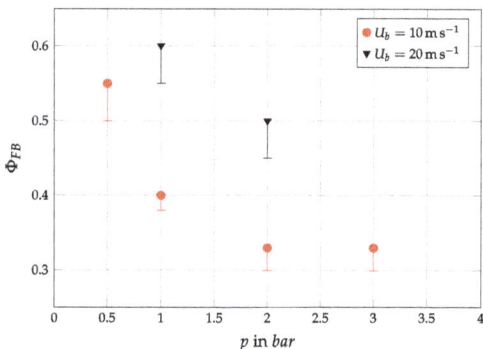

Figure 5. Pressure influence on numerical flashback limits.

3.2. Macroscopic Flame Structure and Turbulent Flame Speed

The propagating flame fronts at all four pressure levels can be compared in Figure 6. It is evident that flame front wrinkling increases with pressure. Furthermore, the radius of the propagating flame tongue decreases with increasing pressure. This is caused by the decrease of flame thickness with higher pressure. The smaller flame thickness makes the flame more susceptible to flame wrinkling and allows for smaller flame structures. The flame thickness δ_f is defined by

$$\delta_f = \frac{T_{ad} - T_u}{\max\left(\left|\frac{\partial T}{\partial x_i}\right|\right)}, \tag{13}$$

where $\partial T / \partial x_i$ is the temperature gradient. The turbulent flame thickness is obtained by sampling the magnitude of the temperature gradient of one time step on isosurfaces of the progress variable c. The progress variable for lean mixtures is defined by the unburnt hydrogen mass fraction $Y_{H_2,u}$ and the local hydrogen mass fraction Y_{H_2}:

$$c = \frac{Y_{H_2,u} - Y_{H_2}}{Y_{H_2,u}}. \tag{14}$$

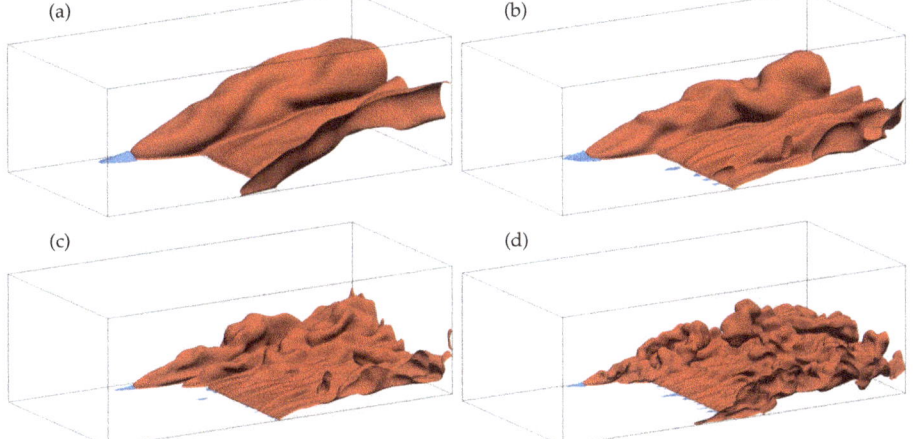

Figure 6. Instantaneous flame fronts and separation zones during flashback at the pressure levels (**a**) 0.5 bar, (**b**) 1 bar, (**c**) 2 bar and (**d**) 3 bar for $U_b = 10\,\mathrm{m\,s^{-1}}$. The flame front is represented by the red $c = 0.5$ isosurface. The separation zone is represented by the blue $u = 0$ isosurface.

19 isosurfaces of c between 0.05 and 0.95 from at least 27 time steps during flame flashback are considered for sampling the temperature gradient. Wall normal turbulent flame thickness profiles are obtained for each of the 19 isosurfaces by calculating the flame thickness from the streamwise, spanwise and time average of the sampled temperature gradient. A minimum average flame thickness of all flame thickness profiles is computed for each wall normal distance. This approach results in the minimum flame thickness profiles in Figure 7. It is evident that although the equivalence ratio is reduced with increasing pressure, the flame thickness significantly reduces with increasing pressure.

The flame thickness profiles in Figure 7 can be used to obtain a minimum resolution of the simulations. The flame thickness is therefore divided by the cube root of the cell volume. This results in minimum resolutions of 3.0 cells per flame thickness for $U_b = 10\,\mathrm{m\,s^{-1}}$ and $\Phi = 0.33$ at $p = 3\,\mathrm{bar}$ up to 5.5 cells per flame thickness for $U_b = 10\,\mathrm{m\,s^{-1}}$ and $\Phi = 0.55$ at $p = 0.5\,\mathrm{bar}$. The minimum resolution requirement of 3 cells per flame thickness [25] is thus fulfilled for all cases.

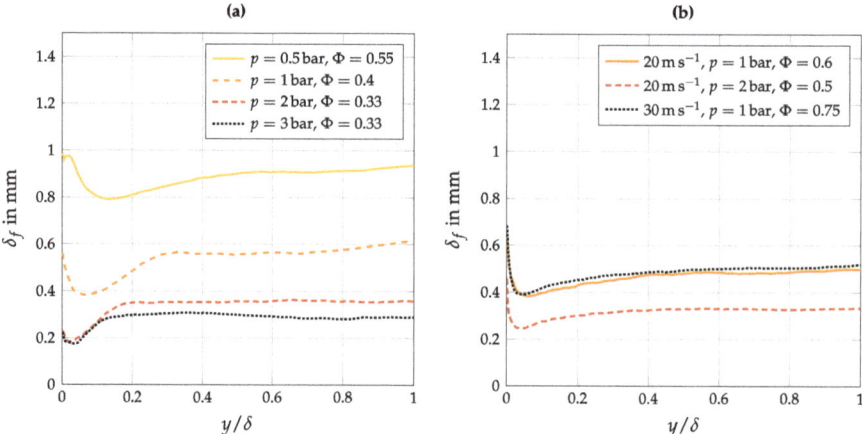

Figure 7. Flame thickness profiles at flashback conditions for (a) $U_b = 10\,\text{m s}^{-1}$ and for (b) $U_b = 20\,\text{m s}^{-1}$ and $U_b = 30\,\text{m s}^{-1}$.

Increasing the pressure from 2 bar to 3 bar at $U_b = 10\,\text{m s}^{-1}$ and $\Phi = 0.33$ does not decrease the average flame thickness as strongly as at lower pressures. This is confirmed by the laminar flame thicknesses in Figure 8 obtained from freely propagating flame simulations with Cantera [38]. It is evident that at low pressures, flame thickness decreases with increasing pressure. At higher pressures, the flame thickness however increases with increasing pressure. At very high pressure levels, the flame thickness of all equivalence ratios decreases again with increasing pressure. It is therefore expected that the flame thickness is increased and the susceptibility to flame front wrinkling is decreased at low equivalence ratios and higher pressures. In addition, the laminar flame speed decreases with increasing pressure [38]. The turbulent flame speed at flashback conditions does therefore not necessarily increase with pressure.

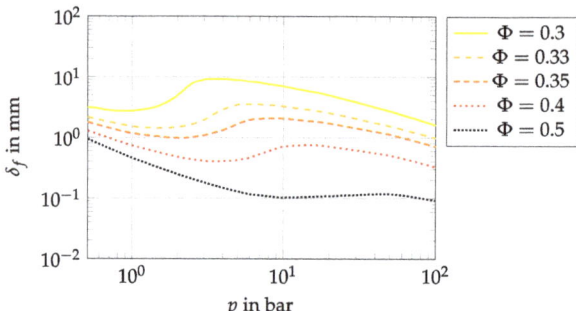

Figure 8. Pressure dependence of the laminar flame thickness obtained from one-dimensional freely propagating flame simulations.

The turbulent flame speed in the simulations can be estimated from Figure 9, where the average stable flame shapes are shown for all operating conditions. The flame shapes are represented by isolines of the averaged progress variable, which are obtained by spatially averaging the temporal average of the hydrogen mass fraction field. The decreasing laminar flame speed at higher pressures can not be compensated by the increased flame front wrinkling due to smaller flame thicknesses. This leads to smaller average flame angles at higher pressure. The average flame shape is however only an inaccurate measure for the turbulent flame thickness. Figure 10 therefore shows the turbulent flame speed profiles of all stable cases at $U_b = 10\,\text{m s}^{-1}$ and $U_b = 20\,\text{m s}^{-1}$. Here, the turbulent flame speed is defined as the average gas velocity component normal to the flame front [40]. This gas velocity is measured at the $\bar{c} = 0.05$ isoline of the average flame front. The measured gas velocity is corrected by the factor T_u/T in order to account for the temperature rise to the flame temperature T at $\bar{c} = 0.05$. The actual turbulent flame speeds depend on the chosen value of $\bar{c}$ for the flame speed evaluation. The turbulent flame speed profiles in Figure 10 are therefore normalized by the maximum turbulent flame speed at the corresponding bulk velocity. The turbulent flame speed profiles demonstrate that the maximum turbulent flame speed at conditions close to flashback varies strongly with pressure. The maximum turbulent flame speed at $U_b = 10\,\text{m}\,\text{s}^{-1}$ and $p = 3\,\text{bar}$ is only 73 % of the maximum turbulent flame speed at $U_b = 10\,\text{m}\,\text{s}^{-1}$ and $p = 0.5\,\text{bar}$. From Equation (4), this difference in turbulent flame speed would result in a deviation of 40 % in flashback channel centerline velocity. This shows that the average flame shape, the turbulent flame speed and Equation (4) are poor indicators for the susceptibility of boundary layer flashback.

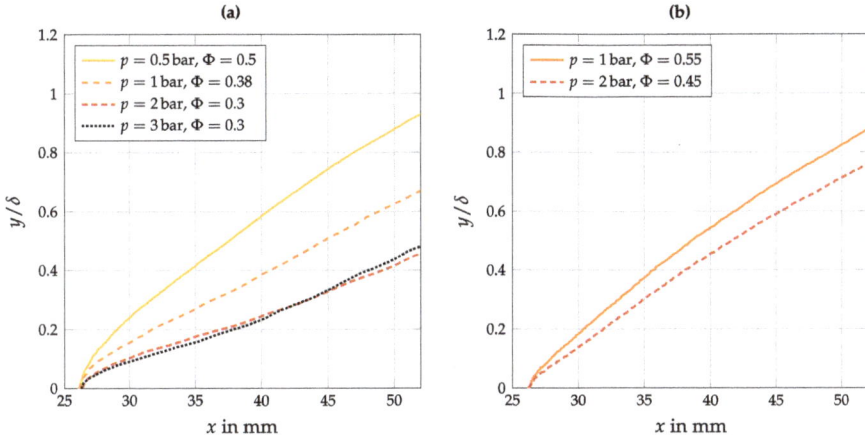

Figure 9. Comparison of average stable flame shapes at different pressure levels evaluated at $\bar{c} = 0.5$. (a) for $U_b = 10\,\text{m}\,\text{s}^{-1}$ and (b) for $U_b = 20\,\text{m}\,\text{s}^{-1}$.

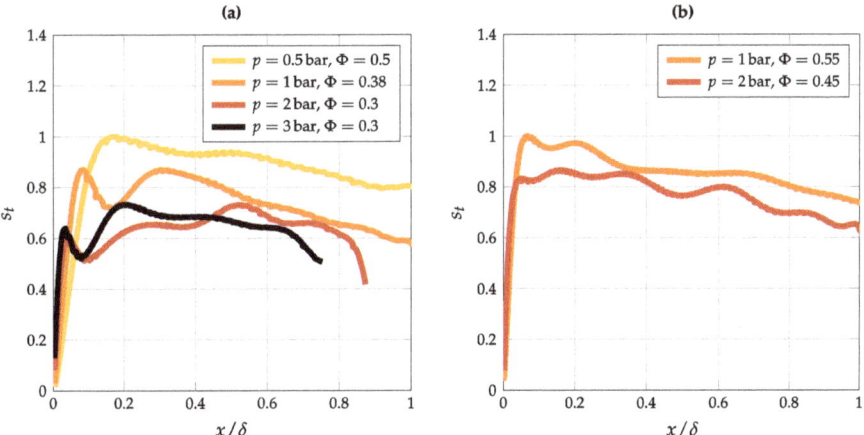

Figure 10. Comparison of turbulent flame speeds at different pressure levels for (a) $U_b = 10\,\text{m}\,\text{s}^{-1}$ and for (b) $U_b = 20\,\text{m}\,\text{s}^{-1}$.

3.3. Average Pressure and Velocity Fields

Eichler et al. [16] showed that confined flames induce a pressure rise ahead of the flame which leads to flow deflection away from the wall. This can also be observed in Figure 11a, where the streamlines of the time and spanwise averaged velocity field are depicted together with the average flame shape and pressure field for the stable case at $U_b = 10\,\text{m}\,\text{s}^{-1}$ and $p = 0.5\,\text{bar}$. Ahead of the flame the streamlines are deflected away from the wall. The turbulent flame brush redirects the streamlines back towards the wall. The flow redirection is caused by the fact that the tangential gas velocity is not changed by the flame front, while the velocity component normal to the flame front increases due to the density decrease across the flame. Ahead of the flame the flow is deflected away from the wall and the axial gas velocity decreases. The two-dimensional flow pattern is linked to a two-dimensional pressure field with a pressure maximum just ahead of the leading flame tip. This confirms the two-dimensional nature of the velocity field and of the pressure rise ahead of the flame observed by Eichler [11].

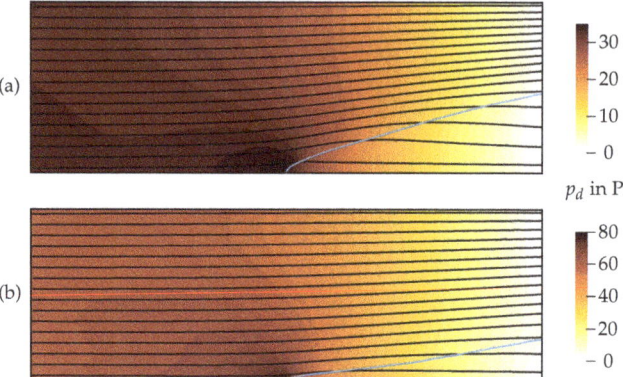

Figure 11. Average streamlines, average flame shape and average pressure field of the stable cases at $U_b = 10\,\text{m}\,\text{s}^{-1}$ at (**a**) $p = 0.5\,\text{bar}$ and at (**b**) $p = 3\,\text{bar}$. The average flame shape is represented by the blue $\bar{c} = 0.5$ isoline.

The flashback equivalence ratio decreases with increasing pressure. The expansion ratio ρ_u/ρ_b of unburnt gas density ρ_u to burnt gas density ρ_b therefore also decreases with increasing pressure. This leads to weaker flow deflection at higher pressures, which can be seen when comparing the streamlines in Figure 11b for the stable case at $p = 3\,\text{bar}$ with the streamlines in Figure 11a at $p = 0.5\,\text{bar}$. The different streamline patterns in Figure 11 show that average boundary layer separation is not an indicator for flame flashback. The boundary layer at $p = 0.5\,\text{bar}$ shows stronger flow deflection and is much closer to average boundary layer separation than the boundary layer at $p = 3\,\text{bar}$. At elevated pressures, flashback thus occurs at reduced turbulent flame speeds, reduced expansion ratios and at reduced average flow deflection ahead of the flame. These average quantities alone can therefore not describe the flashback propensity of a flame. Instead, local processes at the flame tip, such as flame quenching and local flow separation have to be taken into account.

3.4. Quenching Distance and Local Flow Separation

In addition to the average quantities previously presented, the local quantities at the propagating flame front also show significant differences depending on the operating conditions. By tracking the flame front during flashback, the separation zone size and quenching distance during flashback are obtained. Figure 12 exemplarily shows the hydrogen reaction rate and flow separation zone on a cutting plane through the foremost flame tip during flashback at $U_b = 20\,\text{m}\,\text{s}^{-1}$ and $p = 1\,\text{bar}$. From this cutting plane, a wall-normal profile of the maximum hydrogen reaction rate can be obtained. The quenching distance δ_q is then defined as the wall distance of the first local maximum of this maximum hydrogen reaction rate profile. The separation zone size y_s is the wall-normal size of the region, where the streamwise velocity $u \leq 0$.

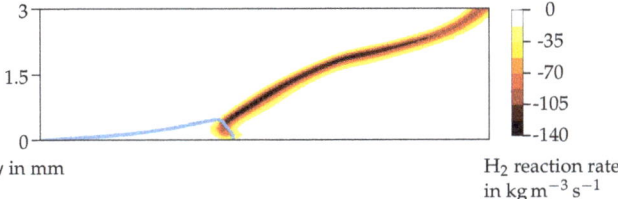

Figure 12. Hydrogen reaction rate and flame separation during flashback. Blue line represents the $u = 0$ isosurface.

Endres and Sattelmayer [15] showed that upstream flame propagation takes place when the separation zone size is significantly larger than the average quenching distance. This is also evident for $U_b = 20\,\text{m}\,\text{s}^{-1}$ and $p = 1\,\text{bar}$ in Figure 13, where the x position of the foremost flame tip is compared to the separation zone size and the quenching distance during flashback. This is however not necessarily always the case. During flashback at $U_b = 10\,\text{m}\,\text{s}^{-1}$ and $p = 0.5\,\text{bar}$, upstream flame propagation also occurs in time periods where the separation zone size is considerably smaller than the quenching distance. This implies that the flame speed for this operating condition is high enough for the flame to propagate upstream at wall distances above the separation zone and not only in the backflow region of the separation zone.

The average quenching distances and separation zone sizes during upstream propagation of all cases under investigation are listed in Table 2. The quenching distance decreases with increasing pressure. This is in accordance with the trend of the flame thickness, which also decreases with pressure. Dividing the quenching distance by the average flame thickness at the quenching wall distance from the profiles in Figure 7 results in the quenching Peclet numbers Pe_q in Table 2. For head-on quenching of laminar hydrogen-oxygen and hydrogen-air flames, Dabireau et al. [44] and Gruber et al. [45] found quenching Peclet numbers of $Pe_q = 1.7$ and $Pe_q = 1.4$. Except for $U_b = 20\,\text{m}\,\text{s}^{-1}$ and $p = 2\,\text{bar}$, where a few cells lie in the thin reaction zones regime, the current quenching Peclet numbers are of the same order of magnitude as the Peclet numbers from literature. The difference in quenching Peclet numbers are likely to arise due to the difference in quenching configuration. The Peclet numbers given here are for turbulent sidewall quenching while the cited literature values are for laminar head-on quenching.

Table 2. Quenching and separation parameters at flashback conditions.

p in bar	U_b in m s^{-1}	Φ	δ_q in mm	$\delta_f\vert_{y=\delta_q}$ in mm	Pe_q	$\overline{y_{sep}}$ in mm	$\overline{y_{sep}}/\delta_q$
0.5	10	0.55	0.78	0.81	0.96	0.66	0.84
1	10	0.4	0.42	0.39	1.06	0.65	1.56
1	20	0.6	0.36	0.39	0.92	0.55	1.53
1	30	0.75	0.37	0.39	0.94	0.53	1.45
2	10	0.33	0.24	0.18	1.31	0.54	2.23
2	20	0.5	0.18	0.26	0.68	0.28	1.58
3	10	0.33	0.18	0.18	1.00	0.39	2.16

The average separation zone size also decreases with increasing pressure. With increasing pressure, the flame thickness and the flame tip radius decrease while the fresh gas density increases with increasing pressure. This leads to weaker flow deflection ahead of the flame, as already seen from the average streamlines in Figure 11. From Table 2 it is however evident that the ratio of separation zone size to quenching distance is not constant. For $U_b = 10\,\text{m}\,\text{s}^{-1}$ for example, it increases from 0.84 at $p = 0.5\,\text{bar}$ to 2.23 at $p = 2\,\text{bar}$. The decrease of quenching thickness with pressure is thus stronger than the decrease of separation zone size. There is hence no unique ratio between separation zone size and quenching distance which could be used as a criterion for the onset of flame flashback.

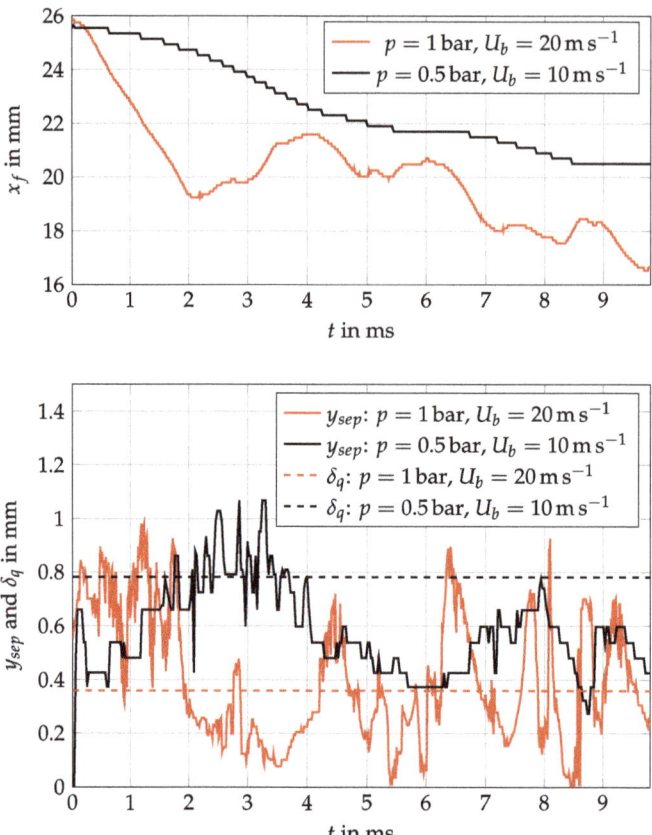

Figure 13. Flame front position, separation zone size and quenching distance during flame flashback. Flame front position x_f is given by the foremost cell with $c > 0.5$.

From $p = 2$ bar to $p = 3$ bar, the ratio of separation zone size to quenching distance is reduced from 2.23 to 2.16. This indicates that here, the separation zone size decreases more strongly than the quenching distance. In addition to the turbulent flame speed, this is another reason why the simulation at $U_b = 10\,\text{m}\,\text{s}^{-1}$ and $p = 3$ bar does not show a higher flashback equivalence ratio than at $p = 2$ bar. It was shown in Section 3.2 that the laminar flame thickness increases with pressure at very lean gas mixtures and higher pressures than investigated in the present study. It is possible that the quenching distance and the separation zone size show a similar trend and increase with increasing pressure. Depending on the ratio of separation zone size to quenching distance, this may increase or decrease the flashback propensity of the confined flames at higher pressures.

3.5. Implications for Analytical Flashback Prediction

The comparison of the pressure rise estimated by the analytical model of Hoferichter et al. [9] with the pressure rise obtained from LES shows that the analytical model does not accurately predict the quantitative values of the pressure rise. According to the analytical model the pressure rise for the stable atmospheric case at $U_b = 10\,\text{m}\,\text{s}^{-1}$ and $\Phi = 0.38$ is 32.2 Pa, while in the simulations only 5.5 Pa are observed. This pressure rise is similar to the pressure rise observed with the compressible solver from Reference [15]. The low Mach number assumption thus does not have a strong effect on the average pressure rise ahead of the flame front. At $U_b = 20\,\mathrm{m\,s^{-1}}$ and $\Phi = 0.55$ and at $U_b = 30\,\mathrm{m\,s^{-1}}$ and $\Phi = 0.7$ the pressure rise ahead of the flame is 22.9 Pa and 63.7 Pa in the simulations. The analytical model however predicts a pressure rise of 106.4 Pa and 214.3 Pa for the same operating conditions. The pressure rise ahead of the flames of all cases is thus overestimated by the one-dimensional pressure calculation of the analytical model. At the same time, the Stratford separation criterion underestimates the separation propensity at BLF. The Stratford criterion is based on the boundary layer theory. The application of the boundary layer theory requires a uniform wall-normal pressure. Figure 14 shows the average wall-normal pressure profile just upstream of the stable atmospheric flame at $U_b = 20\,\mathrm{m\,s^{-1}}$. It is evident that the pressure is not constant over the channel height. Such a non-uniform pressure profile does not only cause flow deceleration but also flow deflection. This increases the tendency for boundary layer separation. The boundary layer theory and the Stratford criterion cannot be applied in this case.

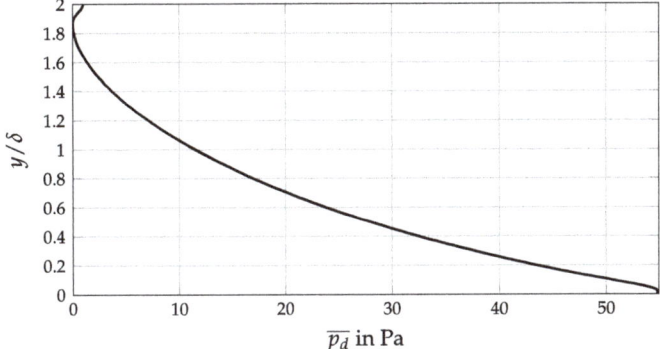

Figure 14. Wall-normal average dynamic pressure profile ahead of the stable flame front at atmospheric pressure and $U_b = 20\,\mathrm{m\,s^{-1}}$.

4. Conclusions

The influence of pressure on the combustion and flow characteristics was investigated by comparing LES results of confined BLF at different pressure levels and bulk velocities. At elevated pressure the flame thickness decreases and the flame is corrugated more strongly by turbulence and thermal-diffusive instabilities. The turbulent flame speed and the flame angle at conditions close to flashback however decreased with increasing pressure. The average flow deflection ahead of the flame also decreased with increasing pressure. Lower turbulent flame speeds and weaker flow deflection ahead of the flame at higher pressure levels would imply a reduced flashback propensity at higher pressure levels. It was however shown that the flashback propensity increased with increasing pressure. The turbulent flame speed, the average flame shape and flow deflection alone are thus poor indicators for the onset of BLF.

Instead, additional quantities such as the quenching distance and the local separation zone size have to be accounted for. The quenching thickness decreases with increasing pressure. At the same time, the local separation zone size ahead of the propagating flame tip decreases with increasing pressure. The reduced flame thickness and quenching distance lead to a smaller flame tip diameter, which reduces the deflection of the approaching flow by the flame. Although these two local effects are counteracting, they do not cancel out and their influence on the flashback limits cannot be neglected. Confined BLF is instead a complex process, which is influenced by the local flame speed, the flame thickness, the quenching distance and the separation tendency of the approaching flow. The low separation zone sizes and high quenching distances at low pressure increase the flashback resistance of the flame, thus allowing for higher stable equivalence ratios. As the flame speed increases with increasing equivalence ratio at low pressures, the susceptibility of the flame to flashback increases. Usually flame propagation occurs inside the recirculating flow of the separation zone. The high flame speeds at low pressures however also allows for flame propagation in regions without recirculating flow.

Finally it was noted that the underlying assumptions of the boundary layer theory and of one-dimensional pressure approximations are not satisfied in confined BLF. The application of one-dimensional pressure approximations leads to an overestimation of the pressure rise ahead of the flame. The application of the boundary layer theory leads to an underestimation of the separation probability. These two errors are counteracting and might possibly compensate each other. Due to this inconsistency, the analytical flashback prediction model [9] should however rather be regarded as a correlation on the experimental flashback data, which is capable of accurately accounting for flashback influences such as preheat temperature or fuel composition.

Author Contributions: Conceptualization: A.E. and T.S.; methodology, software, validation, formal analysis, investigation, writing—original draft preparation, visualization: A.E.; writing—review and editing, supervision: T.S.

Funding: This research was funded by the German research foundation DFG, project No. SA 781/19-1.

Acknowledgments: The authors gratefully acknowledge the compute and data resources provided by the Leibniz Supercomputing Centre (www.lrz.de).

Conflicts of Interest: The authors declare no conflict of interest. The funders had no role in the design of the study; in the collection, analyses, or interpretation of data; in the writing of the manuscript, or in the decision to publish the results.

Abbreviations

The following abbreviations are used in this manuscript:

BLF   Boundary Layer Flashback
LES   Large Eddy Simulation
ILES   Implicit LEL
ODE   Ordinary Differential Equations
SGS   Subgrid Scale

Appendix A

Table A1. Parameters for the mixture diffusivity calculation.

p in bar	Φ	T_{ad} in K	$s_{l,0}\|_{p=1\,\text{bar}}$ in m s^{-1}	m	α in m^2 s^{-1}	ν in m^2 s^{-1}	Le	D in m^2 s^{-1}
0.5	0.55	1741.4	0.76	-0.26	7.275×10^{-5}	3.722×10^{-5}	0.458	1.587×10^{-4}
1	0.4	1422.3	0.35	-0.35	3.248×10^{-5}	1.754×10^{-5}	0.394	8.246×10^{-5}
1	0.6	1838.6	0.92	-0.24	3.698×10^{-5}	1.864×10^{-5}	0.484	7.640×10^{-5}
1	0.75	2096.8	1.40	-0.18	4.003×10^{-5}	1.944×10^{-5}	0.572	7.000×10^{-5}
2	0.33	1257.2	0.20	-0.40	1.559×10^{-5}	8.689×10^{-6}	0.370	4.212×10^{-5}
2	0.5	1640.7	0.61	-0.29	1.763×10^{-5}	9.166×10^{-6}	0.435	4.053×10^{-5}
3	0.33	1257.2	0.20	-0.40	1.039×10^{-5}	5.793×10^{-6}	0.370	2.808×10^{-5}

References

1. Bolland, O.; Undrum, H. A novel methodology for comparing CO2 capture options for natural gas-fired combined cycle plants. *Adv. Environ. Res.* **2003**, *7*, 901–911. [CrossRef]
2. Voldsund, M.; Jordal, K.; Anantharaman, R. Hydrogen production with CO2 capture. *Int. J. Hydrogen Energy* **2016**, *41*, 4969–4992. [CrossRef]
3. Schorr, M.M.; Chalfin, J. *Gas Turbine NOx Emissions Approaching Zero — Is It Worth the Price?* General Electric Power Generation, Report No. GER 4172; General Electric Power: Schenectady, NY, USA, 1999.
4. Gruber, A.; Chen, J.H.; Valiev, D.; Law, C.K. Direct numerical simulation of premixed flame boundary layer flashback in turbulent channel flow. *J. Fluid Mech.* **2012**, *709*, 516–542. [CrossRef]

5. Daniele, S.; Jansohn, P.; Boulouchos, K. Flashback Propensity of Syngas Flames at High Pressure: Diagnostic and Control. In *Combustion, Fuels and Emissions, Parts A and B*; ASME: Glasgow, UK, 2010; Volume 2, pp. 1169–1175, doi:10.1115/GT2010-23456.
6. Kalantari, A.; Sullivan-Lewis, E.; McDonell, V. Flashback Propensity of Turbulent Hydrogen–Air Jet Flames at Gas Turbine Premixer Conditions. *J. Eng. Gas Turbines Power* **2016**, *138*, 061506. [CrossRef]
7. Eichler, C.; Sattelmayer, T. Experiments on flame flashback in a quasi-2D turbulent wall boundary layer for premixed methane-hydrogen-air mixtures. *J. Eng. Gas Turbines Power* **2011**, *133*, 011503. [CrossRef]
8. Hoferichter, V.; Sattelmayer, T. Boundary Layer Flashback in Premixed Hydrogen–Air Flames with Acoustic Excitation. *J. Eng. Gas Turbines Power* **2018**, *140*, 051502. [CrossRef]
9. Hoferichter, V.; Hirsch, C.; Sattelmayer, T. Prediction of Confined Flame Flashback Limits Using Boundary Layer Separation Theory. *J. Eng. Gas Turbines Power* **2017**, *139*, 021505. [CrossRef]
10. Stratford, B.S. The prediction of separation of the turbulent boundary layer. *J. Fluid Mech.* **1959**, *5*, 1–16. [CrossRef]
11. Eichler, C.T. *Flame Flashback in Wall Boundary Layers of Premixed Combustion Systems*; Verlag Dr. Hut: München, Germany, 2011.
12. Gruber, A.; Richardson, E.S.; Aditya, K.; Chen, J.H. Direct numerical simulations of premixed and stratified flame propagation in turbulent channel flow. *Phys. Rev. Fluids* **2018**, *3*. [CrossRef]
13. Clemens, N. *Large Eddy Simulation Modeling of Flashback and Flame Stabilization in Hydrogen-Rich Gas Turbines Using a Hierarchical Validation Approach*; University of Texas: Austin, TX, USA, 2015; doi:10.2172/1253136.
14. Lietz, C.; Hassanaly, M.; Raman, V.; Kolla, H.; Chen, J.; Gruber, A. LES of Premixed Flame Flashback in a Turbulent Channel. In Proceedings of the 52nd Aerospace Sciences Meeting, National Harbor, MD, USA, 13–17 January 2014; American Institute of Aeronautics and Astronautics: Reston, VA, USA, 2014; doi:10.2514/6.2014-0824. [CrossRef]
15. Endres, A.; Sattelmayer, T. Large Eddy simulation of confined turbulent boundary layer flashback of premixed hydrogen-air flames. *Int. J. Heat Fluid Flow* **2018**, *72*, 151–160. [CrossRef]
16. Eichler, C.; Baumgartner, G.; Sattelmayer, T. Experimental Investigation of Turbulent Boundary Layer Flashback Limits for Premixed Hydrogen-Air Flames Confined in Ducts. *J. Eng. Gas Turb. Power* **2012**, *134*, 011502. [CrossRef]
17. Weller, H.G.; Tabor, G.; Jasak, H.; Fureby, C. A tensorial approach to computational continuum mechanics using object-oriented techniques. *Comput. Phys.* **1998**, *12*, 620. [CrossRef]
18. Jasak, H. Error Analysis and Estimation for Finite Volume Method with Applications to Fluid Flow. Ph.D. Thesis, University of London, London, UK, June 1996.
19. Nozaki, F. Smagorinsky SGS Model in OpenFOAM. Available online: https://caefn.com/openfoam/smagorinsky-sgs-model (accessed on 1 June 2019).
20. Smagorinsky, J. General circulation experiments with the primitive equations: I. The basic experiment. *Mon. Weather Rev.* **1963**, *91*, 99–164. [CrossRef]
21. Lilly, D.K. The representation of small scale turbulence in numerical simulation experiments. In Proceedings of the IBM Scientific Computing Symposium on Environmental Sciences, Yorktown Heights, NY, USA, 14–16 November 1966; pp. 195–210.
22. van Driest, E.R. On turbulent flow near a wall. *J. Aerosp. Sci.* **1956**, *23*, 1007–1011. [CrossRef]
23. Mukha, T.; Liefvendahl, M. *Large-Eddy Simulation of Turbulent Channel Flow*; Technical Report Number 2015-014; Uppsala University: Uppsala, Sweden, 2015.
24. Pope, S.B. *Turbulent Flows*; Cambridge University Press: Cambridge, MA, USA, 2000.
25. Duwig, C.; Nogenmyr, K.J.; Chan, C.k.; Dunn, M.J. Large Eddy Simulations of a piloted lean premix jet flame using finite-rate chemistry. *Combust. Theory Model.* **2011**, *15*, 537–568. [CrossRef]
26. Hairer, E.; Wanner, G. *Solving Ordinary Differential Equations II*; Springer Series in Computational Mathematics; Springer: Berlin/Heidelberg, Germany, 1996; Volume 14, doi:10.1007/978-3-642-05221-7.
27. Burke, M.P.; Chaos, M.; Ju, Y.; Dryer, F.L.; Klippenstein, S.J. Comprehensive H2/O2 kinetic model for high-pressure combustion. *Int. J. Chem. Kinet.* **2012**, *44*, 444–474. [CrossRef]
28. Contino, F.; Jeanmart, H.; Lucchini, T.; D'Errico, G. Coupling of in situ adaptive tabulation and dynamic adaptive chemistry: An effective method for solving combustion in engine simulations. *Proc. Combust. Inst.* **2011**, *33*, 3057–3064. [CrossRef]

29. Pope, S. Computationally efficient implementation of combustion chemistry using in situ adaptive tabulation. *Combust. Theory Model.* **1997**, *1*, 41–63. [CrossRef]
30. Krüger, O.; Duwig, C.; Terhaar, S.; Paschereit, C.O. Ultra-Wet Operation of a Hydrogen Fueled GT Combustor: Large Eddy Simulation Employing Detailed Chemistry. In Proceedings of the Seventh International Conference on Computational Fluid Dynamics (ICCFD7), ICCFD7-3403, Big Island, HI, USA, 9–13 July 2012.
31. Fureby, C. Comparison of flamelet and finite rate chemistry LES for premixed turbulent combustion. In Proceedings of the 45th AIAA Aerospace Sciences Meeting and Exhibit, Reno, Nevada, 8–11 January 2007; p. 1413.
32. Pitsch, H.; Duchamp de Lageneste, L. Large-eddy simulation of premixed turbulent combustion using a level-set approach. *Proc. Combust. Inst.* **2002**, *29*, 2001–2008. [CrossRef]
33. Heimerl, J.M.; Coffee, T.P. Transport Algorithms for Methane Flames. *Combust. Sci. Technol.* **1983**, *34*, 31–43. [CrossRef]
34. Kee, R.J.; Rupley, F.M.; Meeks, E.; Miller, J.A. *CHEMKIN-III: A FORTRAN Chemical Kinetics Package for the Analysis of Gas-Phase Chemical and Plasma Kinetics*; Sandia National Laboratories Report SAND96-8216; Sandia National Laboratories: Albuquerque, NW, USA, 1996.
35. Hirschfelder, J.O.; Bird, R.B.; Spotz, E.L. The Transport Properties of Gases and Gaseous Mixtures. II. *Chem. Rev.* **1949**, *44*, 205–231. [CrossRef] [PubMed]
36. Welty, J.R. *Fundamentals of Momentum, Heat, and Mass Transfer*, 5th ed.; Wiley: Hoboken, NJ, USA, 2008.
37. Kuo, K.K.; Acharya, R. *Applications of Turbulent and Multiphase Combustion*; Wiley: Hoboken, NJ, USA, 2012.
38. Goodwin, D.G.; Moffat, H.K.; Speth, R.L. Cantera: An Object-oriented Software Toolkit for Chemical Kinetics, Thermodynamics, and Transport Processes. 2017, doi:10.5281/zenodo.170284. Available online: http://www.cantera.org/ (accessed on 1 June 2019).
39. Sutherland, W. LII. The viscosity of gases and molecular force. *Philos. Mag. Ser. 5* **1893**, *36*, 507–531. [CrossRef]
40. Peters, N. *Turbulent Combustion*; Cambridge Monographs on Mechanics, Cambridge University Press: Cambridge, UK; New York, NY, USA, 2000.
41. Böck, L.R. Deflagration-to-Detonation Transition and Detonation Propagation in H2-Air Mixtures with Transverse Concentration Gradients. Ph.D. Thesis, Technische Universität München, München, Germany, 2015.
42. Bradley, D.; Lawes, M.; Liu, K.; Verhelst, S.; Woolley, R. Laminar burning velocities of lean hydrogen–air mixtures at pressures up to 1.0 MPa. *Combust. Flame* **2007**, *149*, 162–172. [CrossRef]
43. Bechtold, J.K.; Matalon, M. The dependence of the Markstein length on stoichiometry. *Combust. Flame* **2001**, *127*, 1906–1913. [CrossRef]
44. Dabireau, F.; Cuenot, B.; Vermorel, O.; Poinsot, T. Interaction of flames of H2 + O2 with inert walls. *Combust. Flame* **2003**, *135*, 123–133. [CrossRef]
45. Gruber, A.; Sankaran, R.; Hawkes, E.R.; Chen, J.H. Turbulent flame–wall interaction: A direct numerical simulation study. *J. Fluid Mech.* **2010**, *658*, 5–32. [CrossRef]

© 2019 by the authors. Licensee MDPI, Basel, Switzerland. This article is an open access article distributed under the terms and conditions of the Creative Commons Attribution (CC BY) license (http://creativecommons.org/licenses/by/4.0/).

Article

Closure Relations for Fluxes of Flame Surface Density and Scalar Dissipation Rate in Turbulent Premixed Flames

Andrei N. Lipatnikov [1,*], Shinnosuke Nishiki [2] and Tatsuya Hasegawa [3]

1. Department of Mechanics and Maritime Sciences, Chalmers University of Technology, 412 96 Gothenburg, Sweden
2. Department of Mechanical Engineering, Kagoshima University, Kagoshima 890-0065, Japan; nishiki@mech.kagoshima-u.ac.jp
3. Institute of Materials and Systems for Sustainability, Nagoya University, Nagoya 464-8603, Japan; t-hasegawa@imass.nagoya-u.ac.jp
* Correspondence: andrei.lipatnikov@chalmers.se; Tel.: +46-31-772-1386

Received: 11 February 2019; Accepted: 4 March 2019; Published: 7 March 2019

Abstract: In this study, closure relations for total and turbulent convection fluxes of flame surface density and scalar dissipation rate were developed (i) by placing the focus of consideration on the flow velocity conditioned to the instantaneous flame within the mean flame brush and (ii) by considering the limiting behavior of this velocity at the leading and trailing edges of the flame brush. The model was tested against direct numerical simulation (DNS) data obtained from three statistically stationary, one-dimensional, planar, premixed turbulent flames associated with the flamelet regime of turbulent burning. While turbulent fluxes of flame surface density and scalar dissipation rate, obtained in the DNSs, showed the countergradient behavior, the model predicted the total fluxes reasonably well without using any tuning parameter. The model predictions were also compared with results computed using an alternative closure relation for the flame-conditioned velocity.

Keywords: turbulent flame; premixed turbulent combustion; countergradient transport; flame surface density; scalar dissipation rate; modeling; direct numerical simulations

---

1. Introduction

Among various approaches to modeling premixed turbulent combustion, concepts that deal with a transport equation for the mean flame surface density (FSD) $\langle \Sigma \rangle = \langle |\nabla c| \rangle$ [1,2] or the Favre-averaged scalar dissipation rate (SDR) $\langle \chi \rangle^\sim = \langle \rho D \nabla c \cdot \nabla c \rangle / \langle \rho \rangle$ [3] have become particularly popular over the past decade [4–6]. Here, c is the combustion progress variable, which characterizes the state of a reacting mixture in a flame, and is equal to zero and unity in unburned reactants and equilibrium combustion products, respectively, ρ is the mixture density, D is the molecular diffusivity of c, and $\langle q \rangle$ and $\langle q \rangle^\sim = \langle \rho q \rangle / \langle \rho \rangle$ designate the Reynolds and Favre (i.e., mass-weighted), respectively, mean values of an arbitrary quantity q. As reviewed elsewhere [4–6], both transport equations involve a number of terms that should be modeled, but the present communication is solely restricted to total transport terms $\langle \mathbf{u}\Sigma \rangle = \langle \mathbf{u} \rangle \langle \Sigma \rangle + \langle \mathbf{u}'\Sigma' \rangle$ and $\langle \rho \mathbf{u}\chi \rangle = \langle \rho \rangle \langle \mathbf{u} \rangle^\sim \langle \chi \rangle^\sim + \langle \rho \mathbf{u}''\chi'' \rangle$, turbulent contributions to which (i.e., $\langle \mathbf{u}'\Sigma' \rangle$ and $\langle \rho \mathbf{u}''\chi'' \rangle$) require closure relations. Here, u is the flow velocity vector, and $q' = q - \langle q \rangle$ and $q'' = q - \langle q \rangle^\sim$ are fluctuations of q with respect to its Reynolds and Favre-averaged values, respectively.

As reviewed elsewhere [4–6], in applications, the turbulent transport terms are commonly modeled invoking a paradigm of gradient diffusion, for example, $\langle \mathbf{u}'\Sigma' \rangle = -D_t \nabla \langle \Sigma \rangle$ and $\langle \rho \mathbf{u}''\chi'' \rangle = -\langle \rho \rangle D_t \nabla \langle \chi \rangle^\sim$, where $D_t > 0$ is turbulent diffusivity. However, at least for the turbulent flux $\langle \rho \mathbf{u}''c'' \rangle$

of the combustion progress variable c, such a paradigm is challenged by countergradient turbulent transport [7,8], that is, $\langle \rho \mathbf{u}''c'' \rangle \cdot \nabla \langle c \rangle^{\sim} > 0$, which is well documented in various flames, as reviewed elsewhere [9–11]. For the FSD or SDR concept, an issue of eventually countergradient flux of $\langle \mathbf{u}'\Sigma' \rangle$ or $\langle \rho \mathbf{u}''\chi'' \rangle$ is of particular importance, because the corner-stone assumption of the two concepts (i.e., a linear relation between the mean mass rate $\langle W \rangle$ of creation of c and $\rho_u \langle \Sigma \rangle$ [1,2] or $\langle \rho \rangle \langle \chi \rangle^{\sim}$ [3]) is best justified under conditions of the flamelet combustion regime [12], which are commonly associated with the countergradient transport [9–11]. Accordingly, the goal of the present work is to suggest and validate simple closure relations capable of predicting the total convection fluxes $\langle \mathbf{u}\Sigma \rangle = \langle \mathbf{u} \rangle \langle \Sigma \rangle + \langle \mathbf{u}'\Sigma' \rangle$ and $\langle \rho \mathbf{u}\chi \rangle = \langle \rho \rangle \langle \mathbf{u} \rangle^{\sim} \langle \chi \rangle^{\sim} + \langle \rho \mathbf{u}''\chi'' \rangle$ and, in particular, the countergradient turbulent fluxes $\langle \mathbf{u}'\Sigma' \rangle$ and $\langle \rho \mathbf{u}''\chi'' \rangle$ in the flamelet combustion regime.

In the next section, the developed model is presented and the direct numerical simulation (DNS) data used to access it are summarized. In the third section, results of the model validation using the DNS data are discussed and the model is compared with alternative approaches. Conclusions are drawn in the fourth section.

2. Method of Research

2.1. Model

Using velocities

$$\langle \mathbf{u} \rangle_{f,\Sigma} = \langle \mathbf{u}\Sigma \rangle / \langle \Sigma \rangle \tag{1}$$

and

$$\langle \mathbf{u} \rangle_{f,\chi} = \langle \rho \mathbf{u}\chi \rangle / \langle \rho\chi \rangle \tag{2}$$

conditioned to flamelets, modeling of the total fluxes $\langle \mathbf{u}\Sigma \rangle = \langle \mathbf{u} \rangle_{f,\Sigma} \langle \Sigma \rangle$ and $\langle \rho \mathbf{u}\chi \rangle = \langle \mathbf{u} \rangle_{f,\chi} \langle \rho \rangle \langle \chi \rangle^{\sim}$ in the transport equations for $\langle \Sigma \rangle$ and $\langle \chi \rangle^{\sim}$, respectively, is reduced to modeling the conditioned velocities, whereas closure relations for the turbulent fluxes $\langle \mathbf{u}'\Sigma' \rangle$ and $\langle \rho \mathbf{u}''\chi'' \rangle$ are not required.

In order to model the conditioned velocities, let us start with a discussion of a constant-density flow. In such a case, the following simple closure relation,

$$\langle \mathbf{u} \rangle_f = (1 - \langle c \rangle)\langle \mathbf{u} \rangle_b + \langle c \rangle \langle \mathbf{u} \rangle_u, \tag{3}$$

was proposed [13] and validated in a recent DNS study [14] of self-propagation of an infinitely thin interface in a constant-density turbulent flow. Here, $\langle \mathbf{u} \rangle_f$ is velocity conditioned to the interface, subscripts u and b designate quantities conditioned to reactants and products, respectively, which are separated by the interface. Note that the conditioned velocities $\langle \mathbf{u} \rangle_u$ and $\langle \mathbf{u} \rangle_b$ vary along the normal to the mean reaction wave. For instance, in the statistically 1D, planar, and stationary case, $\langle \mathbf{u} \rangle_u = \langle \mathbf{u} \rangle_u(x)$ and $\langle \mathbf{u} \rangle_b = \langle \mathbf{u} \rangle_b(x)$ if the x-axis is normal to the mean wave, with averaging being performed over transverse planes.

Equation (3) is based on the following simple physical reasoning. For an infinitely thin interface, its arrival to the trailing edge of a mean reaction-wave brush is always accompanied by the arrival of reactants to the same trailing point. Therefore, $\langle \mathbf{u} \rangle_f \rightarrow \langle \mathbf{u} \rangle_u$ at $\langle c \rangle \rightarrow 1$. Due to similar arguments, $\langle \mathbf{u} \rangle_f \rightarrow \langle \mathbf{u} \rangle_b$ at $\langle c \rangle \rightarrow 0$. Thus, Equation (3) is nothing more than a linear interpolation between two limiting relations.

Let us assume that the same interpolation holds both for $\langle \mathbf{u} \rangle_{f,\Sigma}$ and $\langle \mathbf{u} \rangle_{f,\chi}$, that is, $\langle \mathbf{u} \rangle_{f,\Sigma} = \langle \mathbf{u} \rangle_{f,\chi} = \langle \mathbf{u} \rangle_f$, in the case of a constant-density "flamelet" of a finite thickness. Such an assumption is based on a small thickness of a flamelet when compared to a mean flame brush thickness in a typical case.

If we allow for combustion-induced thermal expansion, effects due to the finite flamelet thickness appear to be of more importance due to significant velocity variations within thin flamelets. In particular, $\langle \mathbf{u} \rangle_{f,b} \cdot \mathbf{n} < \langle \mathbf{u} \rangle_f \cdot \mathbf{n} < \langle \mathbf{u} \rangle_{f,u} \cdot \mathbf{n}$, because local flow acceleration within a flamelet occurs in the direction opposite to the direction of a unit vector $\mathbf{n} = -\nabla c / |\nabla c|$ that is locally normal to the flamelet (i.e., $(\langle \mathbf{u} \rangle_{f,b} - \langle \mathbf{u} \rangle_{f,u}) \cdot \mathbf{n} < 0$). Here, $\langle \mathbf{u} \rangle_{f,u}$ and $\langle \mathbf{u} \rangle_{f,b}$ are evaluated on the reactant and product sides of the flamelet along the local normal to it. Because the peak value of $\Sigma(c)$ or $\chi(c)$ is shifted to the product side of a typical laminar premixed flame, the simplest way to allow for the discussed difference between the conditioned velocities $\langle \mathbf{u} \rangle_{f,u}$, $\langle \mathbf{u} \rangle_f$, and $\langle \mathbf{u} \rangle_{f,b}$ appears to consist in (i) neglecting the difference in $\langle \mathbf{u} \rangle_{f,\Sigma}$ or $\langle \mathbf{u} \rangle_{f,\chi}$ and $\langle \mathbf{u} \rangle_{f,b} \to \langle \mathbf{u} \rangle_b$ at $\langle c \rangle \to 0$, but (ii) assuming that both $|\langle \mathbf{u} \rangle_{f,b} - \langle \mathbf{u} \rangle_{f,\Sigma}|$ $\ll |\langle \mathbf{u} \rangle_{f,b} - \langle \mathbf{u} \rangle_{f,u}|$ and $|\langle \mathbf{u} \rangle_{f,b} - \langle \mathbf{u} \rangle_{f,\chi}| \ll |\langle \mathbf{u} \rangle_{f,b} - \langle \mathbf{u} \rangle_{f,u}|$ at $\langle c \rangle \to 1$. For example,

$$(\langle \mathbf{u} \rangle_{f,b} - \langle \mathbf{u} \rangle_{f,\Sigma}) = \sigma^{-1}(\langle \mathbf{u} \rangle_{f,b} - \langle \mathbf{u} \rangle_{f,u}) \text{ or } (\langle \mathbf{u} \rangle_{f,b} - \langle \mathbf{u} \rangle_{f,\chi}) = \sigma^{-1}(\langle \mathbf{u} \rangle_{f,b} - \langle \mathbf{u} \rangle_{f,u})$$

in order to avoid invoking tuning parameters. Here, $\sigma = \rho_u / \rho_b$ is the density ratio. Consequently,

$$\langle \mathbf{u} \rangle_{f,\Sigma} \to \sigma^{-1} \langle \mathbf{u} \rangle_{f,u} + (1 - \sigma^{-1}) \langle \mathbf{u} \rangle_{f,b} \text{ and } \langle \mathbf{u} \rangle_{f,\chi} \to \sigma^{-1} \langle \mathbf{u} \rangle_{f,u} + (1 - \sigma^{-1}) \langle \mathbf{u} \rangle_{f,b} \text{ at } \langle c \rangle \to 1.$$

If we further assume that $\langle \mathbf{u} \rangle_{f,b} = \langle \mathbf{u} \rangle_b$ at $\langle c \rangle \to 1$, then,

$$\langle \mathbf{u} \rangle_{f,\Sigma} \to \sigma^{-1} \langle \mathbf{u} \rangle_u + (1 - \sigma^{-1}) \langle \mathbf{u} \rangle_b \text{ and } \langle \mathbf{u} \rangle_{f,\chi} \to \sigma^{-1} \langle \mathbf{u} \rangle_u + (1 - \sigma^{-1}) \langle \mathbf{u} \rangle_b \text{ at } \langle c \rangle \to 1.$$

Note that $\langle \mathbf{u} \rangle_{f,u} \to \langle \mathbf{u} \rangle_u$ at $\langle c \rangle \to 1$, because the unburned gas can arrive at the trailing edge of the mean flame brush only together with a flamelet. Finally, since $\langle \mathbf{u} \rangle_b = \langle \mathbf{u} \rangle = \langle \mathbf{u} \rangle^\sim$ at $\langle c \rangle = 1$, we arrive at

$$\langle \mathbf{u} \rangle_{f,\Sigma} \to \sigma^{-1} \langle \mathbf{u} \rangle_u + (1 - \sigma^{-1}) \langle \mathbf{u} \rangle^\sim \text{ and } \langle \mathbf{u} \rangle_{f,\chi} \to \sigma^{-1} \langle \mathbf{u} \rangle_u + (1 - \sigma^{-1}) \langle \mathbf{u} \rangle^\sim \text{ at } \langle c \rangle \to 1$$

or

$$(\langle \mathbf{u} \rangle_{f,\Sigma} - \langle \mathbf{u} \rangle^\sim) \to \sigma^{-1}(\langle \mathbf{u} \rangle_u - \langle \mathbf{u} \rangle^\sim) \text{ and } (\langle \mathbf{u} \rangle_{f,\chi} - \langle \mathbf{u} \rangle^\sim) \to \sigma^{-1}(\langle \mathbf{u} \rangle_u - \langle \mathbf{u} \rangle^\sim) \text{ at } \langle c \rangle \to 1.$$

Thus, in the case of $\sigma > 1$, we have the following two approximate limiting relations: $\langle \mathbf{u} \rangle_{f,\Sigma} \to \langle \mathbf{u} \rangle_b$ or $\langle \mathbf{u} \rangle_{f,\chi} \to \langle \mathbf{u} \rangle_b$ at $\langle c \rangle \to 0$ and $(\langle \mathbf{u} \rangle_{f,\Sigma} - \langle \mathbf{u} \rangle^\sim) \to \sigma^{-1}(\langle \mathbf{u} \rangle_u - \langle \mathbf{u} \rangle^\sim)$ or $(\langle \mathbf{u} \rangle_{f,\chi} - \langle \mathbf{u} \rangle^\sim) \to \sigma^{-1}(\langle \mathbf{u} \rangle_u - \langle \mathbf{u} \rangle^\sim)$ at $\langle c \rangle \to 1$. Then, linear interpolation results in

$$\langle \mathbf{u} \rangle_{f,\Sigma} - \langle \mathbf{u} \rangle^\sim = \langle c \rangle \sigma^{-1}(\langle \mathbf{u} \rangle_u - \langle \mathbf{u} \rangle^\sim) + (1 - \langle c \rangle)(\langle \mathbf{u} \rangle_b - \langle \mathbf{u} \rangle^\sim) \qquad (4)$$

and

$$\langle \mathbf{u} \rangle_{f,\chi} - \langle \mathbf{u} \rangle^\sim = \langle c \rangle \sigma^{-1}(\langle \mathbf{u} \rangle_u - \langle \mathbf{u} \rangle^\sim) + (1 - \langle c \rangle)(\langle \mathbf{u} \rangle_b - \langle \mathbf{u} \rangle^\sim). \qquad (5)$$

In the case of $\sigma = 1$ (constant density), Equations (4) and (5) reduce to Equation (3) and yield correct values of $\langle \mathbf{u} \rangle_{f,\Sigma}$ and $\langle \mathbf{u} \rangle_{f,\chi}$ at least at the boundaries $\langle c \rangle \to 0$ and $\langle c \rangle \to 1$ of the mean "flame" brush.

It is worth noting that Equations (4) and (5) extend Equation (3) by allowing for effects due to density variations and finite flamelet thickness. Since such an extension was performed by invoking a few simple assumptions, the developed model required validation and the results of such a validation performed by analyzing DNS data will be reported in Section 3. As we will see in that section, the model works reasonably well, but some differences between the model predictions and the DNS data were observed, probably due to the invoked simplifications. It is also worth remembering that the conditioned velocities $\langle \mathbf{u} \rangle_u$ and $\langle \mathbf{u} \rangle_b$ in Equations (4) and (5) are not averaged over the flame brush volume, but vary along the normal to the mean flame brush.

Finally, using the following well-known Bray–Moss–Libby (BML) expressions [8,9,12],

$$\rho_u (1 - \langle c \rangle) = \langle \rho \rangle (1 - \langle c \rangle^\sim), \quad \rho_b \langle c \rangle = \langle \rho \rangle \langle c \rangle^\sim, \qquad (6)$$

$$\langle \mathbf{u} \rangle^\sim = (1 - \langle c \rangle^\sim) \langle \mathbf{u} \rangle_u + \langle c \rangle^\sim \langle \mathbf{u} \rangle_b, \qquad (7)$$

and

$$\langle \rho \mathbf{u}'' c'' \rangle = \langle \rho \rangle \langle c \rangle^\sim (1 - \langle c \rangle^\sim)(\langle \mathbf{u} \rangle_b - \langle \mathbf{u} \rangle_u), \qquad (8)$$

we arrive at

$$\begin{aligned}\langle \mathbf{u}\rangle_{f,\Sigma} = \langle \mathbf{u}\rangle_{f,X} &= \langle \mathbf{u}\rangle^{\sim} - \langle c\rangle\sigma^{-1}\langle\rho\mathbf{u}''c''\rangle/[\langle\rho\rangle(1-\langle c\rangle^{\sim})] + (1-\langle c\rangle)\langle\rho\mathbf{u}''c''\rangle/[\langle\rho\rangle\langle c\rangle^{\sim}]\\ &= \langle \mathbf{u}\rangle^{\sim} - \langle c\rangle^{\sim}\langle\rho\mathbf{u}''c''\rangle''/[\rho_u(1-\langle c\rangle^{\sim})] + \langle\rho\mathbf{u}''c''\rangle(1-\langle c\rangle^{\sim})/[\rho_u\langle c\rangle^{\sim}]\\ &= \langle \mathbf{u}\rangle^{\sim} + (1-2\langle c\rangle^{\sim})\langle\rho\mathbf{u}''c''\rangle/[\rho_u\langle c\rangle^{\sim}(1-\langle c\rangle^{\sim})]. \end{aligned} \quad (9)$$

The problem of modeling the turbulent flux $\langle\rho\mathbf{u}''c''\rangle$ in Equation (9) is not specific to the FSD or SDR concept and should be resolved by any approach that deals with the transport equation for the Favre-averaged $\langle c\rangle^{\sim}$. Accordingly, there are various closure relations for $\langle\rho\mathbf{u}''c''\rangle$, as reviewed elsewhere [9–11]. Therefore, modeling of this flux is beyond the scope of the present communication, that is, $\langle\rho\mathbf{u}''c''\rangle$ is considered to be known here. Thus, Equation (9) is the final result of the present model. It is worth stressing that this closure relation does not involve any tuning parameter.

2.2. Direct Numerical Simulations

In order to test Equation (9), we analyzed DNS data obtained earlier by Nishiki et al. [15,16]. Because these data were used by different research groups in a number of papers [17–36], we will restrict ourselves to a brief summary of these compressible simulations.

The well-known unsteady 3D balance equations for mass, momentum, energy, and mass fraction Y of the deficient reactant were numerically solved. The ideal gas state equation was used. Combustion chemistry was reduced to a single reaction. The molecular transport coefficients were increased by the temperature T (e.g., the kinematic viscosity $\nu = \nu_u(T/T_u)^{0.7}$). The Lewis and Prandtl numbers were equal to 1.0 and 0.7, respectively. Accordingly, the mixture state was completely characterized with a single combustion progress variable $c = 1 - Y/Y_u = (T - T_u)/(T_b - T_u)$.

The computational domain was a rectangular box $\Lambda_x \times \Lambda_y \times \Lambda_z$, with $\Lambda_x = 8$ mm and $\Lambda_y = \Lambda_z = 4$ mm, and was resolved using a uniform rectangular ($2\Delta x = \Delta y = \Delta z$) mesh of $512 \times 128 \times 128$ points. The x-axis was normal to the mean flame brush and was parallel to the direction of its propagation.

Homogeneous isotropic turbulence (the rms turbulent velocity $u' = 0.53$ m/s, an integral length scale of turbulence $L = 3.5$ mm, the turbulent Reynolds number $Re_t = u'L/\nu_u = 96$ [15]) was generated in a separate box and was injected into the computational domain through the left boundary $x = 0$. In the computational domain, the turbulence decayed along the direction x of the mean flow. The flow was periodic in y and z directions.

At $t = 0$, a planar laminar flame ($c = 0$ at $x = 0$ and $c = 1$ at $x = \Lambda_x$) was embedded into statistically the same turbulence assigned for the velocity field in the entire computational domain. Subsequently, in order to keep the flame in the computational domain until the end t_3 of the simulations, the mean inflow velocity, which was parallel to the x-axis, was increased twice, that is, $U(0 \leq t < t_1) = S_L < U(t_1 \leq t < t_2) < U(t_2 \leq t)$, with $U(t_2 \leq t)$ being approximately equal to the mean turbulent flame speed $\langle S_T\rangle$. Here, S_L is the laminar flame speed.

Three cases signifying high (H), medium (M), and low (L) respectively, density ratios were simulated. The basic flame characteristics are reported in Table 1, where $Ka_{th} = (u'/S_L)^{3/2}(L/\delta_{th})^{-1/2}$ and $Da_{th} = (S_L/u')(L/\delta_{th})$ are the Karlovitz and Damköhler numbers, respectively, $\delta_{th} = (T_b - T_u)/\max|\nabla T|$ is the thermal thickness of the laminar flame, and the mean turbulent burning velocity is equal to

$$\rho_u\langle U_T\rangle = [\Lambda_y\Lambda_z(t_3 - t_2)]^{-1}\int\int\int\int W(\mathbf{x},t)dxdt, \quad (10)$$

with the mean turbulent flame speed $\langle S_T\rangle$ being equal to $\langle U_T\rangle$ in the considered statistically one-dimensional, planar case.

Table 1. Studied cases. H–high, M–medium, L–low.

Case	σ	Re$_t$	u'/S$_L$	L$_{11}$/δ$_F$	Da	Ka	⟨U$_T$⟩/S$_L$
H	7.53	96	0.88	15.9	18.1	0.24	1.91
M	5.0	96	1.0	18.0	18.0	0.24	1.90
L	2.50	96	1.26	21.8	17.3	0.24	1.89

As discussed in detail elsewhere [25], the present DNS conditions and results are fully consistent with the paradigm of the flamelet regime [12] of premixed turbulent combustion. Accordingly, the DNS data are particularly useful for the goal of the present communication. Indeed, (i) the DNS conditions and data are consistent with the BML, FSD, and SDR concepts, see References [25,29], and [25], respectively, (ii) the axial turbulent flux ⟨ρu"c"⟩ shows the countergradient behavior [16] due to combustion-induced thermal expansion effects [9–11], and (iii) the use of different density ratios offers an opportunity to vary the magnitude of these effects.

Results presented in the next section were averaged over transverse planes and over time $t_2 \leq t < t_3$ (approximately 200 snapshots). During that time interval, the computed turbulent burning velocity and flame brush thickness oscillated around statistically steady values [26].

3. Results and Discussion

Results of the developed simple model validation are reported in Figures 1–6. Dotted lines in Figure 1a, Figure 2a, and Figure 3a show the total axial FSD flux ⟨uΣ⟩ extracted directly from the DNS data, whereas solid lines show the flux ⟨u⟩$_{f,Σ}$⟨Σ⟩ computed using the model, that is, Equations (1) and (9), with ⟨c⟩~, ⟨u⟩~, ⟨ρu"c"⟩, and ⟨Σ⟩ being extracted from the same DNS data. In all three cases H, M, and L, the agreement between the DNS and model results is very good. Similarly, Figure 1b, Figure 2b, and Figure 3b show the total axial SDR flux ⟨ρuχ⟩ extracted directly from the DNS data, whereas solid lines show the flux ⟨u⟩$_{f,χ}$⟨ρχ⟩ computed using the model, that is, Equations (2) and (9), with ⟨c⟩~, ⟨u⟩~, ⟨ρu"c"⟩, and ⟨ρχ⟩ being extracted from the same DNS data. Again, the model is well validated in all three cases.

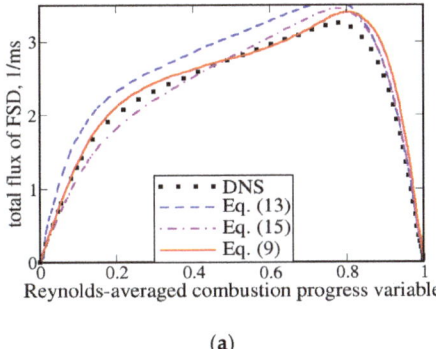
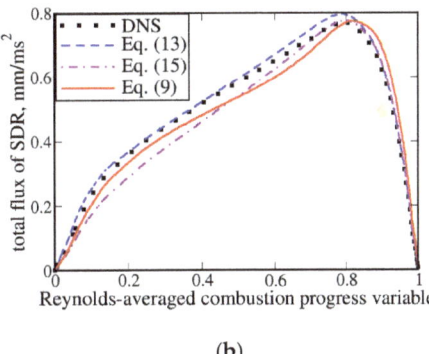

(a)   (b)

Figure 1. Total axial fluxes of (**a**) flame surface density (FSD) Σ and (**b**) scalar dissipation rate (SDR) χ. Case H.

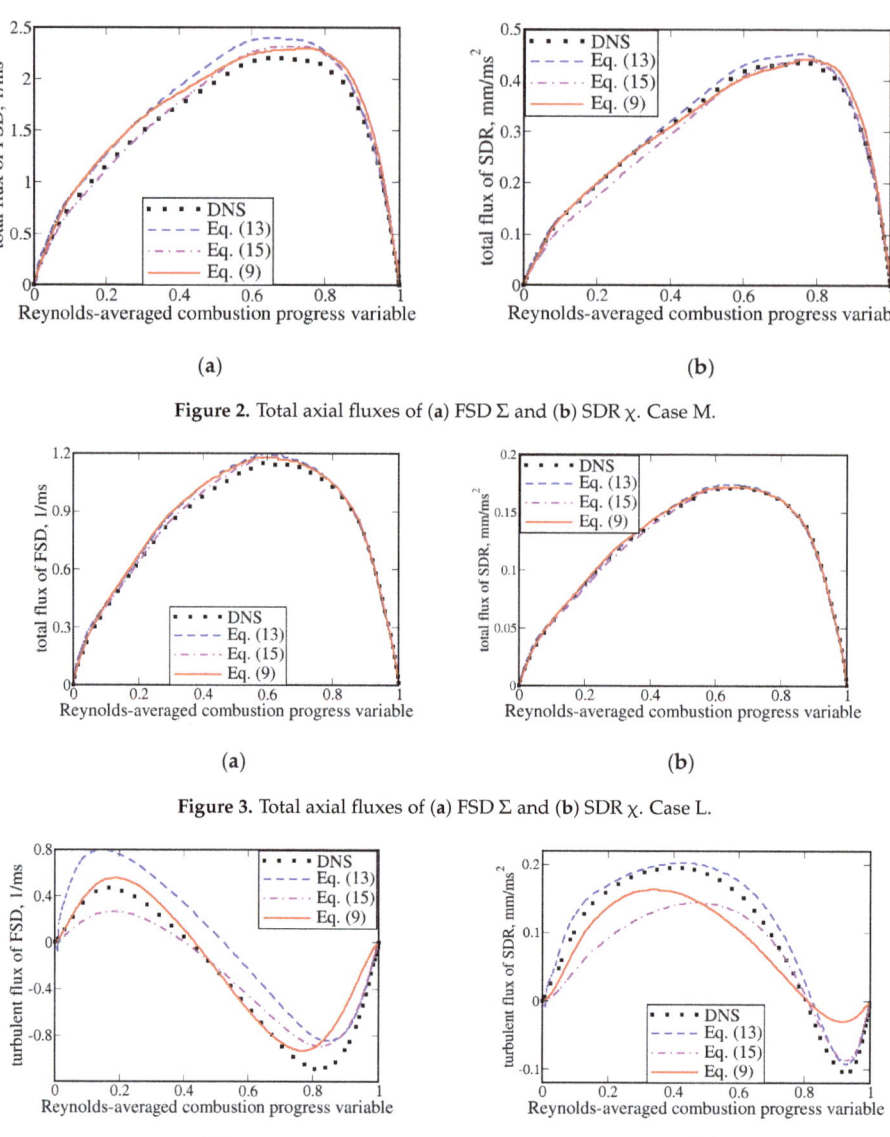

Figure 2. Total axial fluxes of (**a**) FSD Σ and (**b**) SDR χ. Case M.

Figure 3. Total axial fluxes of (**a**) FSD Σ and (**b**) SDR χ. Case L.

Figure 4. Turbulent axial fluxes of (**a**) FSD Σ and (**b**) SDR χ. Case H.

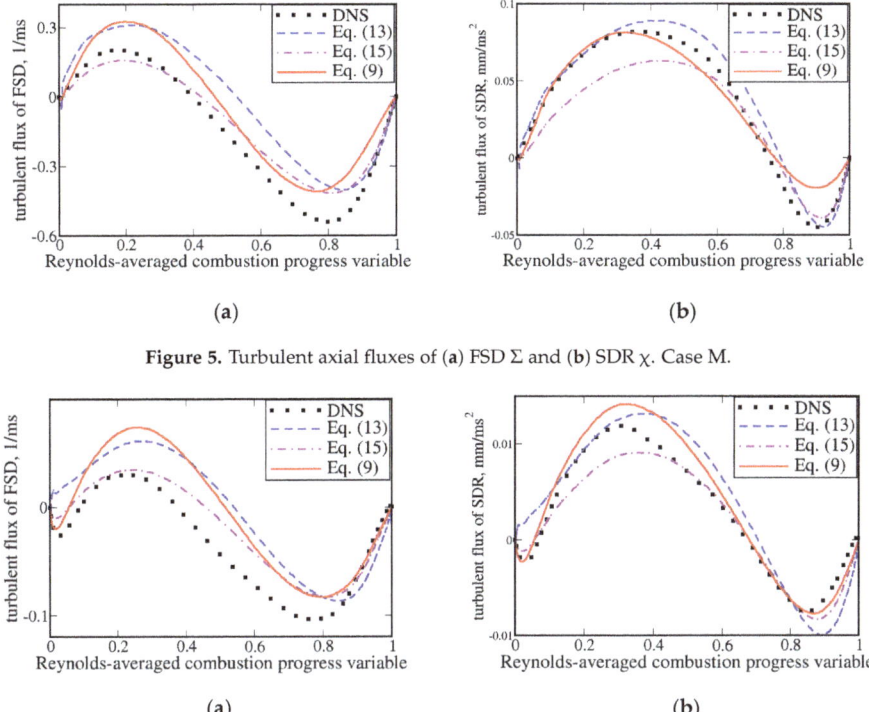

Figure 5. Turbulent axial fluxes of (**a**) FSD Σ and (**b**) SDR χ. Case M.

Figure 6. Turbulent axial fluxes of (**a**) FSD Σ and (**b**) SDR χ. Case L.

It is worth noting, however, that the magnitudes of the total fluxes $\langle \mathbf{u}\Sigma \rangle$ and $\langle \rho \mathbf{u} \chi \rangle$ can mainly be controlled by the mean advection fluxes $\langle \mathbf{u} \rangle \langle \Sigma \rangle$ and $\langle \rho \rangle \langle \mathbf{u} \rangle^{\sim} \langle \chi \rangle^{\sim}$, respectively, which do not require modeling. Accordingly, to better access Equation (9), turbulent fluxes were also evaluated as follows:

$$\langle \mathbf{u}'\Sigma' \rangle = \langle \mathbf{u}\Sigma \rangle - \langle \mathbf{u} \rangle \langle \Sigma \rangle, \tag{11}$$

$$\langle \rho \mathbf{u}'' \chi'' \rangle = \langle \rho \mathbf{u} \chi \rangle - \langle \rho \rangle \langle \mathbf{u} \rangle^{\sim} \langle \chi \rangle^{\sim}. \tag{12}$$

Results plotted in Figures 4–6 show that Equation (9) predicts the axial turbulent fluxes $\langle \mathbf{u}'\Sigma' \rangle$ and $\langle \rho \mathbf{u}'' \chi'' \rangle$ very well in the largest parts of mean flame brushes, that is, at $\langle c \rangle < 0.8$, in cases H and M, respectively, cf. dotted and solid lines in Figures 4a and 5b, respectively. Moreover, the flux $\langle \rho \mathbf{u}'' \chi'' \rangle$ is well predicted in case L (see Figure 6b), whereas the peak value of the flux is slightly overestimated. As far as flux $\langle \mathbf{u}'\Sigma' \rangle$ in cases M and L or flux $\langle \rho \mathbf{u}'' \chi'' \rangle$ in case H is concerned, Equation (9) yields qualitatively correct results, but some quantitative differences between them and the DNS data are observed, probably due to a simplified treatment of the effects of density variations and finite flamelet thickness in Equations (4) and (5). Nevertheless, these differences do not impede quantitatively predicting the total fluxes $\langle \mathbf{u} \rangle \langle \Sigma \rangle$ and $\langle \rho \rangle \langle \mathbf{u} \rangle^{\sim} \langle \chi \rangle^{\sim}$, in particular, because $\langle \mathbf{u}\Sigma \rangle$ is close to the mean flux $\langle \mathbf{u} \rangle \langle \Sigma \rangle$ in cases M and L, cf. scales of ordinate axes in Figure 2a or Figure 3a and in Figure 5a or Figure 6a, respectively.

It is worth stressing that the turbulent axial fluxes $\langle \mathbf{u}'\Sigma' \rangle$ and $\langle \rho \mathbf{u}'' \chi'' \rangle$ show the countergradient behavior in all studied cases (i.e., $\langle \mathbf{u}'\Sigma' \rangle d\langle \Sigma \rangle / dx > 0$ and $\langle \rho \mathbf{u}'' \chi'' \rangle d\langle \rho \chi \rangle / dx > 0$), because $d\langle \Sigma \rangle / dx$ and $d\langle \rho \chi \rangle / dx$ are positive (negative) in the leading (trailing) part of the mean flame brushes (reactants on the left side). Consequently, the widely used concept of gradient diffusion yields wrong direction of the fluxes $\langle \mathbf{u}'\Sigma' \rangle$ and $\langle \rho \mathbf{u}'' \chi'' \rangle$ under conditions of the present DNSs.

It is also worth noting that an alternative model for evaluating the flamelet-conditioned velocity $\langle \mathbf{u} \rangle_f$ was suggested by Bidaux and Bray in an unpublished work (1994) and was subsequently used by Veynante et al. [37]. The model consists in the following linear interpolation:

$$\langle \mathbf{u} \rangle_f = (1 - K)\langle \mathbf{u} \rangle_u + K\langle \mathbf{u} \rangle_b, \tag{13}$$

where the constant K could "be related to the iso-c line used to define the flame location" [37]. Equations (7), (8), and (13) yield

$$\langle \mathbf{u} \rangle_f = \langle \mathbf{u} \rangle^{\sim} + (K - \langle c \rangle^{\sim})\langle \rho \mathbf{u}''c'' \rangle / [\langle \rho \rangle \langle c \rangle^{\sim}(1 - \langle c \rangle^{\sim})]. \tag{14}$$

In the cited paper, the constant K was set equal to 0.5 and we are not aware of any alternative suggestion for K. If K = 0.5, Equation (14) can be rewritten in the following form:

$$\langle \mathbf{u} \rangle_f = \langle \mathbf{u} \rangle^{\sim} + (\rho_u/2\langle \rho \rangle)(1 - 2\langle c \rangle^{\sim})\langle \rho \mathbf{u}''c'' \rangle / [\rho_u \langle c \rangle^{\sim}(1 - \langle c \rangle^{\sim})], \tag{15}$$

which is similar to Equation (9), but the last term on the right-hand side (RHS) of Equation (15) is multiplied by a factor of R = $\rho_u/2\langle \rho \rangle$. This factor tends to 1/2 and $\sigma/2$ at the leading and trailing edges of the mean flame brush. It is equal to unity somewhere in the middle of the mean flame brush.

While Equations (13)–(15) were never applied to model turbulent fluxes of FSD and SDR, to the best of our knowledge, this gap can easily be filled by assuming that $\langle \mathbf{u} \rangle_{f,\Sigma} = \langle \mathbf{u} \rangle_{f,\chi} = \langle \mathbf{u} \rangle_f$ and using Equations (1) or (2), (11) or (12), and (13)–(15). The outcomes of such tests of the model by Bidaux and Bray with K = 0.5 are shown in dashed and dotted-dashed lines in Figures 1–6. There are some differences between the results computed using Equations (13) and (15), because Equation (15) is obtained from Equation (13) using Equations (7) and (8), which are exact only in the case of an infinitely thin flame front. In the present DNSs, flamelets have finite thicknesses and, consequently, there are some (sufficiently small) differences between the left-hand side and right-hand side of Equation (7) or (8), extracted from the DNS data (e.g., see Figure 3b in Reference [25]).

The comparison of Equation (13) with Equation (9) shows that Equation (13) with K = 0.5 yields better results for flux $\langle \rho u''\chi'' \rangle$ in case H (see Figure 4b), but worse results for flux $\langle u'\Sigma' \rangle$ in cases H and M (see Figures 4a and 5a). Similarly, when compared to Equation (9), Equation (15) yields better results for flux $\langle u'\Sigma' \rangle$ in cases M and L (see Figures 5a and 6a), but worse results for both $\langle u'\Sigma' \rangle$ and $\langle \rho u''\chi'' \rangle$ in case H, see Figure 4. Moreover, for both fluxes, Equation (9) is superior to Equation (13) or (15) at the leading edge of the mean flame brush (i.e., at $\langle c \rangle \to 0$). For instance, Equation (15) yields wrong direction of the fluxes at very low $\langle c \rangle$ in cases H and M, whereas Equation (13) underestimates the magnitudes $|\langle u'\Sigma' \rangle|$ and $|\langle \rho u''\chi'' \rangle|$ at very low $\langle c \rangle$ in all cases. This difference between Equation (9) and Equation (13) or (15) appears to be of paramount importance, because the leading edge of premixed turbulent flame brush can play the crucial role in its propagation, as discussed in detail elsewhere [38–45]. Furthermore, Equation (13) or (15) yields the same wrong limit behavior of $\langle \mathbf{u} \rangle_f \to \langle \mathbf{u} \rangle_u/2 + \langle \mathbf{u} \rangle_b/2$, both at the leading and trailing edges of the mean flame brush in the simplest case of the propagation of an infinitely thin, dynamically passive front in the constant-density turbulent flow. On the contrary, in such a case, Equation (9) obtained in the present work (i) reduces to Equation (3), which was well validated in Reference [14], and (ii) yields $\langle \mathbf{u} \rangle_f \to \langle \mathbf{u} \rangle_b$ and $\langle \mathbf{u} \rangle_f \to \langle \mathbf{u} \rangle_u$ at $\langle c \rangle \to 0$ and $\langle c \rangle \to 1$, respectively, in line with the simple physical reasoning discussed at the beginning of Section 2.1.

4. Conclusions

In this study, closure relations for total and turbulent fluxes of flame surface density Σ and scalar dissipation rate χ were developed. The relations did not involve a tuning parameter, very well predicted the total axial fluxes $\langle u\Sigma \rangle$ and $\langle \rho u\chi \rangle$, and reasonably well predicted turbulent axial fluxes $\langle u'\Sigma' \rangle$ and $\langle \rho u''\chi'' \rangle$ in three flames characterized by significantly different density ratios and associated with (i) the flamelet regime of premixed turbulent burning, (ii) the validity of the FSD and SDR concepts, and (iii) the countergradient turbulent fluxes of Σ and χ. In the case of a constant density, the developed closure relation reduced to Equation (3), which was recently validated under conditions of gradient turbulent transport [14].

Author Contributions: Data curation, S.N. and T.H.; Formal analysis, A.N.L.; Methodology, A.N.L.; Investigation, A.N.L.; Software, S.N.; Writing—original draft preparation, A.N.L.; Writing—review and editing, A.N.L., S.N., and T.H.

Acknowledgments: The first author (Andrei N. Lipatnikov) gratefully acknowledges the financial support provided by the Combustion Engine Research Center (CERC).

Conflicts of Interest: The authors declare no conflict of interest.

References

1. Candel, S.; Veynante, D.; Lacas, F.; Maistret, E.; Darabiha, N.; Poinsot, T. Coherent flame model: Applications and recent extensions. In *Advances in Combustion Modeling*; Larrouturou, B.E., Ed.; World Scientific: Singapore, 1990; pp. 19–64.
2. Cant, S.R.; Pope, S.B.; Bray, K.N.C. Modelling of flamelet surface-to-volume ratio in turbulent premixed combustion. *Proc. Combust. Inst.* **1990**, *23*, 809–815. [CrossRef]
3. Borghi, R. Turbulent premixed combustion: Further discussions of the scales of fluctuations. *Combust. Flame* **1990**, *80*, 304–312. [CrossRef]
4. Veynante, D.; Vervisch, L. Turbulent combustion modeling. *Prog. Energy Combust. Sci.* **2002**, *28*, 193–266. [CrossRef]
5. Poinsot, T.; Veynante, D. *Theoretical and Numerical Combustion*, 2nd ed.; Edwards: Philadelphia, PA, USA, 2005.
6. Chakraborty, N.; Champion, M.; Mura, A.; Swaminathan, N. Scalar-dissipation-rate approach. In *Turbulent Premixed Flames*; Swaminathan, N., Bray, K.N.C., Eds.; Cambridge University Press: Cambridge, UK, 2011; pp. 76–102.
7. Moss, J.B. Simultaneous measurements of concentration and velocity in an open premixed turbulent flame. *Combust. Sci. Technol.* **1980**, *22*, 119–129. [CrossRef]
8. Libby, P.A.; Bray, K.N.C. Countergradient diffusion in premixed turbulent flames. *AIAA J.* **1981**, *19*, 205–213. [CrossRef]
9. Bray, K.N.C. Turbulent transport in flames. *Proc. R. Soc. Lond. A* **1995**, *451*, 231–256. [CrossRef]
10. Lipatnikov, A.N.; Chomiak, J. Effects of premixed flames on turbulence and turbulent scalar transport. *Prog. Energy Combust. Sci.* **2010**, *36*, 1–102. [CrossRef]
11. Sabelnikov, V.A.; Lipatnikov, A.N. Recent advances in understanding of thermal expansion effects in premixed turbulent flames. *Annu. Rev. Fluid Mech.* **2017**, *49*, 91–117. [CrossRef]
12. Bray, K.N.C. Turbulent flows with premixed reactants. In *Turbulent Reacting Flows*; Libby, P.A., Williams, F.A., Eds.; Springer-Verlag: Berlin, Germany, 1980; pp. 115–183.
13. Lipatnikov, A.N. Conditionally averaged balance equations for modeling premixed turbulent combustion in flamelet regime. *Combust. Flame* **2008**, *152*, 529–547. [CrossRef]
14. Yu, R.; Bai, X.-S.; Lipatnikov, A.N. A direct numerical simulation study of interface propagation in homogeneous turbulence. *J. Fluid Mech.* **2015**, *772*, 127–164. [CrossRef]
15. Nishiki, S.; Hasegawa, T.; Borghi, R.; Himeno, R. Modeling of flame-generated turbulence based on direct numerical simulation databases. *Proc. Combust. Inst.* **2002**, *29*, 2017–2022. [CrossRef]
16. Nishiki, S.; Hasegawa, T.; Borghi, R.; Himeno, R. Modelling of turbulent scalar flux in turbulent premixed flames based on DNS databases. *Combust. Theory Model.* **2006**, *10*, 39–55. [CrossRef]
17. Im, Y.H.; Huh, K.Y.; Nishiki, S.; Hasegawa, T. Zone conditional assessment of flame-generated turbulence with DNS database of a turbulent premixed flame. *Combust. Flame* **2004**, *137*, 478–488. [CrossRef]
18. Mura, A.; Tsuboi, K.; Hasegawa, T. Modelling of the correlation between velocity and reactive scalar gradients in turbulent premixed flames based on DNS data. *Combust. Theory Model.* **2008**, *12*, 671–698. [CrossRef]
19. Mura, A.; Robin, V.; Champion, M.; Hasegawa, T. Small scale features of velocity and scalar fields in turbulent premixed flames. *Flow Turbul. Combust.* **2009**, *82*, 339–358. [CrossRef]

20. Robin, V.; Mura, A.; Champion, M.; Hasegawa, T. Modeling of the effects of thermal expansion on scalar turbulent fluxes in turbulent premixed flames. *Combust. Sci. Technol.* **2010**, *182*, 449–464. [CrossRef]
21. Robin, V.; Mura, A.; Champion, M. Direct and indirect thermal expansion effects in turbulent premixed flames. *J. Fluid Mech.* **2011**, *689*, 149–182. [CrossRef]
22. Bray, K.N.C.; Champion, M.; Libby, P.A.; Swaminathan, N. Scalar dissipation and mean reaction rates in premixed turbulent combustion. *Combust. Flame* **2011**, *158*, 2017–2022. [CrossRef]
23. Lipatnikov, A.N.; Nishiki, S.; Hasegawa, T. A direct numerical simulation study of vorticity transformation in weakly turbulent premixed flames. *Phys. Fluids* **2014**, *26*, 105104. [CrossRef]
24. Lipatnikov, A.N.; Sabelnikov, V.A.; Nishiki, S.; Hasegawa, T.; Chakraborty, N. DNS assessment of a simple model for evaluating velocity conditioned to unburned gas in premixed turbulent flames. *Flow Turbul. Combust.* **2015**, *94*, 513–526. [CrossRef]
25. Lipatnikov, A.N.; Nishiki, S.; Hasegawa, T. DNS assessment of relation between mean reaction and scalar dissipation rates in the flamelet regime of premixed turbulent combustion. *Combust. Theory Model.* **2015**, *19*, 309–328. [CrossRef]
26. Lipatnikov, A.N.; Chomiak, J.; Sabelnikov, V.A.; Nishiki, S.; Hasegawa, T. Unburned mixture fingers in premixed turbulent flames. *Proc. Combust. Inst.* **2015**, *35*, 1401–1408. [CrossRef]
27. Sabelnikov, V.A.; Lipatnikov, A.N.; Chakraborty, N.; Nishiki, S.; Hasegawa, T. A transport equation for reaction rate in turbulent flows. *Phys. Fluids* **2016**, *28*, 081701. [CrossRef]
28. Sabelnikov, V.A.; Lipatnikov, A.N.; Chakraborty, N.; Nishiki, S.; Hasegawa, T. A balance equation for the mean rate of product creation in premixed turbulent flames. *Proc. Combust. Inst.* **2017**, *36*, 1893–1901. [CrossRef]
29. Lipatnikov, A.N.; Sabelnikov, V.A.; Nishiki, S.; Hasegawa, T. Flamelet perturbations and flame surface density transport in weakly turbulent premixed combustion. *Combust. Theory Model.* **2017**, *21*, 205–227. [CrossRef]
30. Lipatnikov, A.N.; Sabelnikov, V.A.; Chakraborty, N.; Nishiki, S.; Hasegawa, T. A DNS study of closure relations for convection flux term in transport equation for mean reaction rate in turbulent flow. *Flow Turbul. Combust.* **2018**, *100*, 75–92. [CrossRef] [PubMed]
31. Lipatnikov, A.N.; Chomiak, J.; Sabelnikov, V.A.; Nishiki, S.; Hasegawa, T. A DNS study of the physical mechanisms associated with density ratio influence on turbulent burning velocity in premixed flames. *Combust. Theory Model.* **2018**, *22*, 131–155. [CrossRef]
32. Lipatnikov, A.N.; Sabelnikov, V.A.; Nishiki, S.; Hasegawa, T. Combustion-induced local shear layers within premixed flamelets in weakly turbulent flows. *Phys. Fluids* **2018**, *30*, 085101. [CrossRef]
33. Lipatnikov, A.N.; Sabelnikov, V.A.; Nishiki, S.; Hasegawa, T. Does flame-generated vorticity increase turbulent burning velocity? *Phys. Fluids* **2018**, *30*, 081702. [CrossRef]
34. Sabelnikov, V.A.; Lipatnikov, A.N.; Nishiki, S.; Hasegawa, T. Application of conditioned structure functions to exploring influence of premixed combustion on two-point turbulence statistics. *Proc. Combust. Inst.* **2019**, *37*, 2433–2441. [CrossRef]
35. Lipatnikov, A.N.; Nishiki, S.; Hasegawa, T. A DNS assessment of linear relations between filtered reaction rate, flame surface density, and scalar dissipation rate in a weakly turbulent premixed flame. *Combust. Theory Model.* **2019**, *23*. [CrossRef]
36. Sabelnikov, V.A.; Lipatnikov, A.N.; Nishiki, S.; Hasegawa, T. Investigation of the influence of combustion-induced thermal expansion on two-point turbulence statistics using conditioned structure functions. *J. Fluid Mech.* **2019**, in press.
37. Veynante, D.; Trouvé, A.; Bray, K.N.C.; Mantel, T. Gradient and counter-gradient scalar transport in turbulent premixed flames. *J. Fluid Mech.* **1997**, *332*, 263–293. [CrossRef]
38. Kuznetsov, V.R.; Sabelnikov, V.A. *Turbulence and Combustion*; Hemisphere Publishing Corporation: New York, NY, USA, 1990.
39. Lipatnikov, A.N.; Chomiak, J. Molecular transport effects on turbulent flame propagation and structure. *Prog. Energy Combust. Sci.* **2005**, *31*, 1–73. [CrossRef]
40. Lipatnikov, A.N. *Fundamentals of Premixed Turbulent Combustion*; CRC Press: Boca Raton, FL, USA, 2012.
41. Venkateswaran, P.; Marshall, A.; Seitzman, J.; Lieuwen, T. Scaling turbulent flame speeds of negative Markstein length fuel blends using leading points concepts. *Combust. Flame* **2015**, *162*, 375–387. [CrossRef]
42. Sabelnikov, V.A.; Lipatnikov, A.N. Transition from pulled to pushed fronts in premixed turbulent combustion: Theoretical and numerical study. *Combust. Flame* **2015**, *162*, 2893–2903. [CrossRef]

43. Kim, S.H. Leading points and heat release effects in turbulent premixed flames. *Proc. Combust. Inst.* **2017**, *36*, 2017–2024. [CrossRef]
44. Dave, H.L.; Mohan, A.; Chaudhuri, S. Genesis and evolution of premixed flames in turbulence. *Combust. Flame* **2018**, *196*, 386–399. [CrossRef]
45. Lipatnikov, A.N.; Chakraborty, N.; Sabelnikov, V.A. Transport equations for reaction rate in laminar and turbulent premixed flames characterized by non-unity Lewis number. *Int. J. Hydrog. Energy* **2018**, *43*, 21060–21069. [CrossRef]

© 2019 by the authors. Licensee MDPI, Basel, Switzerland. This article is an open access article distributed under the terms and conditions of the Creative Commons Attribution (CC BY) license (http://creativecommons.org/licenses/by/4.0/).

Article

DNS Study of the Bending Effect Due to Smoothing Mechanism

Rixin Yu [1] and Andrei N. Lipatnikov [2,*]

[1] Division of Fluid Mechanics, Lund University, 221 00 Lund, Sweden; rixin.yu@energy.lth.se
[2] Department of Mechanics and Maritime Sciences, Chalmers University of Technology, 412 96 Gothenburg, Sweden
* Correspondence: andrei.lipatnikov@chalmers.se; Tel.: +46-31-772-1386

Received: 28 January 2019; Accepted: 15 February 2019; Published: 19 February 2019

Abstract: Propagation of either an infinitely thin interface or a reaction wave of a nonzero thickness in forced, constant-density, statistically stationary, homogeneous, isotropic turbulence is simulated by solving unsteady 3D Navier–Stokes equations and either a level set (G) or a reaction-diffusion equation, respectively, with all other things being equal. In the case of the interface, the fully developed bulk consumption velocity normalized using the laminar-wave speed S_L depends linearly on the normalized rms velocity u'/S_L. In the case of the reaction wave of a nonzero thickness, dependencies of the normalized bulk consumption velocity on u'/S_L show bending, with the effect being increased by a ratio of the laminar-wave thickness to the turbulence length scale. The obtained bending effect is controlled by a decrease in the rate of an increase δA_F in the reaction-zone-surface area with increasing u'/S_L. In its turn, the bending of the $\delta A_F(u'/S_L)$-curves stems from inefficiency of small-scale turbulent eddies in wrinkling the reaction-zone surface, because such small-scale wrinkles characterized by a high local curvature are smoothed out by molecular transport within the reaction wave.

Keywords: reaction waves; turbulent reacting flows; turbulent consumption velocity; bending effect; reaction surface area; molecular transport; direct numerical simulations

1. Introduction

In turbulent reacting flows, the so-called bending effect consists in decreasing the rate $(dU_T)/du'$ of an increase in turbulent consumption velocity U_T by the rms turbulent velocity u' with increasing u', i.e., the second derivative of the function $U_T(u')$ is negative. As illustrated in Figure 1, due to this effect, the curve plotted in an orange solid line is bent and, at high u', shows significantly lower consumption velocities when compared to the straight dashed red line. Since this basic phenomenon was documented, e.g., in premixed turbulent flames [1,2], it has been challenging the research community and different approaches to explaining and modeling the bending effect have been put forward.

The most recognized approach consists in highlighting the so-called stretch effect, i.e., variations in the local structure of reaction wave (e.g., flame) and the local consumption velocity u_c, caused by turbulent stretching of the wave [3–5]. Here, u_c is a properly normalized rate of production of a major reaction product, integrated along the local normal to a thin reaction zone, which is assumed to be inherently laminar within the framework of the discussed concept. The straightforward manifestation of the stretch effect consists in changing the mean value $\langle u_c \rangle$ of the local consumption velocity with increasing u', followed by eventual local reaction extinction at sufficiently high u'. A decrease in $\langle u_c \rangle$ and the local reaction extinction can also affect the area A_F of the reaction-wave surface, but this manifestation of the discussed mechanism is indirect, i.e., it is a consequence of the dependence of u_c or $\langle u_c \rangle$ on the local stretch rate or u', respectively. According to the theory of stretched laminar premixed flames [6,7], which is well developed in the case of single-step chemistry and the asymptotically high activation energy E of the global combustion reaction, the dependence of u_c on the local stretch rate s is controlled by differences in molecular diffusivities of a fuel, D_F, oxygen, D_O, and the heat diffusivity κ of the mixture. In the equidiffusive ($D_F = D_O = \kappa$) adiabatic case, u_c does not depend on s if E tends to infinity, but u_c can depend on s and the flame extinction by the stretch rate can occur at a finite activation energy [8]. The reader interested in further discussion of the stretch effect in premixed flames is referred to review paper [9].

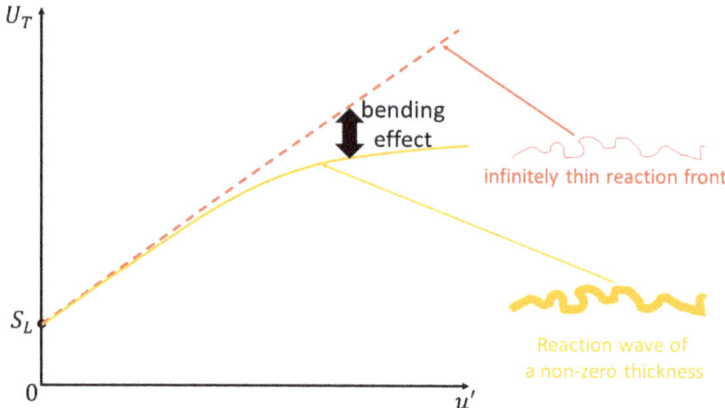

Figure 1. A sketch of the bending effect for turbulent consumption velocity U_T as a function of the rms turbulent velocity u'.

Even if the mean $\langle u_c \rangle$ is close to the speed S_L of the unperturbed laminar reaction wave, the bending effect can occur. For instance, a recent Direct Numerical Simulation (DNS) study [10] has shown that the bending effect can be controlled by the mean flame surface area $\langle A_F \rangle$, i.e., the second derivative of the function $\langle A_F \rangle(u')$ can be negative in spite of $\langle u_c \rangle \approx S_L$. Under such conditions, the bending effect might be attributed to various physical mechanisms, e.g., statistical equilibrium between small-scale turbulent eddies and reaction rate [11], collisions of reaction waves [12,13], or smoothing of small-scale wrinkles of the reaction-wave surface due to its propagation [14].

While all the aforementioned approaches [3–5,11–14] were developed by studying premixed flames, they place the focus of consideration on the influence of turbulence on combustion, but disregard the back influence of the combustion on the turbulence. However, phenomena caused by combustion-induced thermal expansion can also contribute to the bending effect. For instance, small-scale turbulent eddies may be inefficient in wrinkling reaction-zone surface, because they disappear due to dilatation and an increase in the kinematic viscosity of the preheated mixture [15,16]. The reader interested in further discussion of the thermal expansion effects is referred to review papers [17,18].

It is also worth noting that Damköhler [19] has arrived at the following scaling $U_T \propto S_L(u'L/\kappa)^{1/2}$ by reducing the influence of very intense turbulence on reaction wave to enhancement of heat and mass transfer within the wave by turbulent eddies. Here, L is an integral turbulent length scale. From the purely mathematical viewpoint, this scaling results in the bending, but the physical mechanism hypothesized by Damköhler [19] is associated an increase in $\langle u_c \rangle$ when compared to S_L.

Finally, the bending effect can be pronounced differently in different reaction waves. For instance, levelling-off of $U_T(u')$-curves, followed by a decrease in U_T with increasing u', is well documented in expanding statistically spherical flames [1,2], but the positive dU_T/du' was obtained from statistically stationary flames at much higher values of u'/S_L [20,21].

Besides the reaction-wave-configuration effects, which are beyond the scope of the present study, the aforementioned physical mechanisms of the bending may be divided into three groups; (A) mechanisms that highlight differences in the molecular transport coefficients, i.e., the stretch effect [3–5], (B) mechanisms that highlight thermal expansion effects in flames, and (C) mechanisms that address equidiffusive flames, but disregard the thermal expansion effects [11–14,19]. The physical mechanisms of the bending may also be divided into two other groups; (a) mechanisms that highlight variations in local consumption velocity u_c due to the stretch effect [3–5] or the transport enhancement [19] and (b) mechanisms that highlight the bending of $\langle A_F \rangle(u')$-curves.

To conclude this brief introduction, it is worth noting that the fact that various physical mechanisms of the bending effect are discussed in the literature does not mean that all relevant physical mechanisms have already been revealed.

The present communication does not aim at comparing all the aforementioned physical mechanisms. The goals of the communication are solely restricted to (i) supporting recent finding [10] that the bending of $U_T(u')$-curves can be controlled by the bending of $\langle A_F \rangle(u')$-curves, (ii) comparing physical mechanisms from Group (C), and (iii) emphasizing a physical mechanism that controls the bending of $\langle A_F \rangle(u')$-curves under conditions of the present study, but has yet been outside of the mainstream discussions, to the best of the present authors' knowledge. It is worth stressing that, under certain conditions in turbulent flames, the emphasized physical mechanism might play a less important role when compared to preferential diffusion or thermal expansion effects, which are not addressed in the present study. This reservation should be borne in mind when applying the reported results to modeling premixed turbulent combustion.

2. Method of Research

For these purposes, a DNS study of propagation of (i) an infinitely thin interface, see fine red line in Figure 1, and (ii) a single-reaction wave of a finite thickness, see thick orange shape, in statistically the same (in both cases) homogeneous, isotropic, forced, constant-density turbulence affected neither by the interface nor by the reaction was performed.

The constant-density turbulence is described by the continuity

$$\nabla \cdot \mathbf{u} = 0 \tag{1}$$

and Navier–Stokes equations

$$\partial \mathbf{u}/\partial t + (\mathbf{u} \cdot \nabla)\mathbf{u} = -\rho^{-1} \nabla p + \nu \nabla^2 \mathbf{u} + \mathbf{f}, \tag{2}$$

where t is time, $\mathbf{u}$ is the flow velocity vector, ρ, ν, and p are the density, kinematic viscosity, and pressure, respectively, and a vector-function $\mathbf{f}$ is added in order to maintain constant turbulence intensity by using energy forcing at low wavenumbers.

Propagation of the infinitely thin interface is modeled by level set (or G) equation [22]

$$\partial G/\partial t + \mathbf{u} \cdot \nabla G = S_L |\nabla G|, \tag{3}$$

where G is a signed distance function to the closest interface associated with $G(\mathbf{x},t) = 0$. Attributes, methodology, and results of the simulations that dealt with Equations (1)–(3) were already discussed by us in details in References [23,24].

Moreover, propagation of a reaction wave of a non-zero thickness is modeled by the following reaction-diffusion equation

$$\partial c/\partial t + \mathbf{u} \cdot \nabla c = D \Delta c + W, \tag{4}$$

for a scalar field c, which is equal to zero and unity in fresh reactants and products, respectively. Here,

$$W = (1 - c)\exp[-(Ze(1 + \tau)^2)/\tau(1 + \tau c)]/[\tau_r(1 + \tau)] \tag{5}$$

is the reaction rate, $\tau = 6$, and $Ze = 6$. DNS cases are set up (i) by specifying the reaction-wave speed S_L and thickness $\delta_F = D/S_L$ and (ii) by finding required constant values of D and reaction time scale τ_r in pre-simulations of a planar 1D laminar-wave modeled by Equations (4) and (5).

Because the attributes, methodology, and results of the simulations that dealt with Equations (1), (2), (4), and (5) were already discussed by us in detail [25,26], we will restrict ourselves to a very brief summary of those DNSs and refer the interested reader to the cited papers.

The computational domain is a rectangular box of size of $\Lambda_x \times \Lambda \times \Lambda$. It is discretized using a uniform staggered Cartesian grid of $N_x \times N \times N$ cells with $N_x = N(\Lambda_x/\Lambda) = 4N$. Therefore, spatial resolution $\Delta x = \Lambda_x/N_x = \Lambda/N = \Delta y = \Delta z$ is the same in the axial (x) and transverse (y and z) directions. The boundary conditions are periodic in all three directions, thus, enabling a piece of reaction zone that comes to the left boundary (x = 0) at certain t, y and z to enter the computational domain through the right boundary (x = Λ_x) at the same t, y and z, respectively.

Turbulence is generated and forced using techniques discussed elsewhere [27–29]. As shown earlier [23–26], the turbulence achieves statistical stationarity, homogeneity, and isotropy over the entire domain, with correlations $R_{xx}(r) = \langle u(x,y,z,t)u(x+r,y,z,t)\rangle$, $R_{yy}(r) = \langle v(x,y,z,t)v(x,y+r,z,t)\rangle$, and $R_{zz}(r) = \langle w(x,y,z,t)w(x,y,z+r,t)\rangle$ being very close to each over and vanishing at $r = \Lambda/2$. Here, the mean values $\langle \cdot \rangle$ are evaluated by averaging the velocity fields over transverse coordinates and time.

The simulations are performed using three velocity fields A, B, and C, whose characteristics are reported in Table 1. Here, L_{11} is the longitudinal integral length scale of the turbulence evaluated by the integrating the correlation $R_{xx}(r)$ over distance r, $\eta = (\nu^3/\langle\langle\varepsilon\rangle\rangle)^{1/4}$ is the Kolmogorov length scale, $\langle\langle\varepsilon\rangle\rangle$ is the dissipation rate $\varepsilon = 2\nu S_{ij}S_{ij}$ averaged over the computational volume and time, and $S_{ij} = (\partial u_i/\partial x_j + \partial u_j/\partial x_i)/2$ is the rate-of-strain tensor. The major difference between the three velocity fields consists of the width Λ of the computational domain, which controls the length scale L_{11} and the initial Reynolds number $Re_0 = u'\Lambda/(4\nu)$. In other words, L_{11} and Re_0 are increased by increasing Λ, whereas u' or ν remain the same in the simulations.

Table 1. Studied cases.

Case	Re_0	$N_x \times N \times N$	$\eta/\Delta x$	L_{11}/η	u'/S_L	L_{11}/δ_F	Da	Ka	$\langle U_{T,c}\rangle/u'$	$\langle U_{T,G}\rangle/u'$
A1	50	256×64^2	0.68	12	0.5	2.0	4.1	1.34	2.18	3.66
A2	-	-	-	-	1.0	-	2.0	2.69	1.31	2.68
A3	-	-	-	-	2.0	-	1.0	5.38	0.89	2.19
A4	-	-	-	-	5.0	-	0.4	13.4	0.57	1.83
A5	-	-	-	-	10.0	-	0.2	26.9	0.40	1.67
B1	100	512×128^2	0.87	17	0.5	3.7	7.5	0.84	2.45	3.66
B2	-	-	-	-	1.0	-	3.8	1.67	1.69	2.71
B3	-	-	-	-	2.0	-	1.9	3.34	1.27	2.23
B4	-	-	-	-	5.0	-	0.8	8.36	0.80	1.87
B5	-	-	-	-	10.0	-	0.4	16.7	0.55	1.77
C1	200	1024×256^2	1.07	25	0.5	6.7	14.	0.55	2.77	3.70
C2	-	-	-	-	1.0	-	6.7	1.10	2.06	2.80
C3	-	-	-	-	2.0	-	3.4	2.21	1.61	-
C4	-	-	-	-	5.0	-	1.4	5.52	1.11	-
C5	-	-	-	-	10.0	-	0.7	11.0	0.76	1.87
TRW	-	-	-	-	-	2.1	0.2	36.2	0.42	-

All cases studied by solving either Equations (1–3) or Equations (1 and 2) and (4 and 5) and using velocity field A, B, or C are reported in Table 1, where $Da = \tau_t/\tau_f$ and $Ka = \tau_f/\tau_\eta$ are the Damköhler and Karlovitz numbers, respectively, $\tau_f = \delta_F/S_L$, $\tau_t = L_{11}/u'$, and $\tau_\eta = (\nu/\langle\langle\varepsilon\rangle\rangle)^{1/2}$ are the reaction wave, eddy-turn-over, and Kolmogorov time scales, respectively. For each velocity field, five cases characterized by different ratios of u'/S_L are studied by varying S_L. When propagation of a reaction wave of a nonzero thickness is addressed using Equations (1 and 2) and (4 and 5), the laminar-wave thickness δ_F retains the same value in all 15 cases A, B, and C in spite of variations in S_L. In order to keep δ_F constant, the Schmidt number $Sc = \nu/D$ is changed. A ratio of L_{11}/δ_F is increased by increasing the width Λ of the computational domain from field A to field C. In a single TRW (thick reaction wave) case, the ratio of L_{11}/δ_F is decreased by increasing the thickness δ_F when compared to Case C5. It is worth remembering that the values of L_{11}/δ_F, Da, and Ka, reported in Table 1, characterize the reaction waves described with Equations (1 and 2) and (4 and 5), whereas $L_{11}/\delta_F = Da = \infty$ and Ka = 0 for an infinitely thin interface, which is characterized solely with u'/S_L in the present study.

In order to initiate the studied process, either an interface $G(x_0,y,z,t) = 0$ or a planar wave $c(x,t) = c_L(\xi)$ is released at $x_0 = \Lambda_x/2$ and $t = 0$ so that the value of the combustion progress variable integrated over the half-space of $x < x_0$ to be equal to the value of c integrated over the half-space of $x > x_0$. Here, $\xi = x - x_0$ and $c_L(\xi)$ is the pre-computed laminar-wave profile. Subsequently, evolution of the field $G(x,t)$ or $c(x,t)$ is simulated by solving Equation (3) or Equations (4) and (5), respectively. In the former case, $c(x,t) = H[G(x,t)]$, where H is Heaviside function.

Mean bulk turbulent consumption velocity is evaluated as follows [24]

$$\langle U_T \rangle = \Lambda^{-2} d/dt \langle \iiint c(\mathbf{x},t) d\mathbf{x} \rangle \tag{6}$$

because this method can be applied to both sets of simulations. When processing the DNS data obtained by solving the reaction-diffusion Equation (3), the following alternative method

$$\langle U_T \rangle = \Lambda^{-2} \langle \iiint W(\mathbf{x},t) d\mathbf{x} \rangle \tag{7}$$

is also applied. Because Equations (6) and (7) yield very close results in all studied cases, solely values of $\langle U_T \rangle$, obtained using Equation (6), will be reported in the following. Here, $\langle \cdot \rangle$ designates a mean value averaged over time from t^* till $t^* + \Delta t$, the starting instant t^* allows the forced turbulence to reach statistical stationarity, and Δt is longer than 50 time scales $\tau^0 = \Lambda/(4u')$. Reported in the two right columns in Table 1 are normalized fully developed mean bulk turbulent consumption velocities obtained by solving Equations (4 and 5) and (3), respectively. Cases C3 and C4 were not simulated by solving the G-equation (3) in Reference [24].

It is worth stressing that the major difference between the two sets of the simulations consists in solving either the G-equation (3) or the reaction-diffusion Equation (4), whereas numerical methods, turbulence characteristics, etc. are basically similar in both sets. Accordingly, in the present communication, the focus of consideration is placed on phenomena that stem from the finite thickness of the reaction wave.

As far as the physical mechanisms discussed in Section 1 are concerned, the present simulations allow for all mechanisms from Group (C). Moreover, the stretch effect is also addressed. Indeed, variations in the local consumption velocity in and the local quenching of stretched inherently laminar flames can occur in the studied adiabatic equidiffusive flames, because the Zeldovich number Ze in Equation (5) is finite [8]. However, it is worth remembering that the stretch effect can be much more pronounced if $D_F \neq D_O \neq \kappa$ or if heat losses play a substantial role. The thermal expansion effects are not taken into account in the present simulations, because the physical mechanisms from Group (C) do not allow for them. We may also note that (i) the vast majority of approximations of experimental data on U_T, e.g., see review papers [30,31], do not invoke the density ratio σ, thus, implying a weak influence of σ on U_T or S_T, (ii) recent target-directed experiments [32], as well as earlier measurements [33], did not reveal a substantial influence of σ on U_T either, and (iii) recent DNS studies, e.g., Figures 10 and 11 in [34] or Figure 2a in [35], do not indicate such an influence.

3. Results

Normalized fully developed turbulent consumption velocities $\langle U_T \rangle/S_L$ obtained by solving either the G-equation (2) or the reaction-diffusion Equation (3), with all other things being equal, are plotted in open or filled symbols, respectively, in Figure 2. Comparison of the two sets of results indicates, first, that $\langle U_{T,G} \rangle$ is significantly higher than $\langle U_{T,c} \rangle$, with the difference in $\langle U_{T,G} \rangle$ and $\langle U_{T,c} \rangle$ being increased by u'/S_L. Therefore, the nonzero thickness of the reaction wave reduces its bulk consumption velocity when compared to the infinitely thin interface. Second, if u'/S_L is kept constant, then, the ratio of

$\langle U_{T,c}\rangle/S_L$ is increased by L_{11}/δ_F, cf. filled blue squares, black circles, and red triangles. Third, while $\langle U_{T,G}\rangle/S_L$ depends linearly on u'/S_L, see open symbols and dashed lines, the solid curves, which fit to the DNS data on $\langle U_{T,c}\rangle/S_L$, are clearly bent. Fourth, the bending effect is more pronounced at a lower L_{11}/δ_F, see filled blue squares.

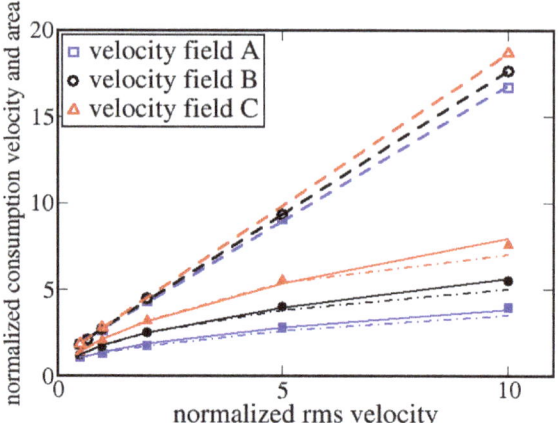

Figure 2. Normalized fully developed turbulent consumption velocity $\langle U_T\rangle/S_L$ (symbols) and a relative increase δA in the wave surface area (dotted-dashed lines) vs normalized root mean square (rms) velocity u'/S_L. Open symbols show DNS data obtained in the case of an infinitely thin interface, while dashed lines are linear fits to these data. Filled symbols show results simulated in the case of a reaction wave of a nonzero thickness, while solid lines are power-law fits to these data.

It is worth noting that the computed increase in $\langle U_{T,c}\rangle/S_L$ by L_{11}/δ_F or u'/S_L agrees with available experimental data, at least qualitatively. In particular, solid lines in Figure 2 show power-law fits $a(u'/S_L)^b$, where b = 0.44, 0.51, and 0.58 for L_{11}/δ_F = 2.0, 3.7, and 6.7, respectively. Similar values of the scaling exponent q in $U_T \propto u'^q$ were documented in various experimental studies, e.g., $q \approx 0.5$ [30,31,36,37], q = 0.56, see data by Kido et al. [38] fitted in Reference [31], q = 0.63 [39], or q = 0.67, see data by Kobayashi et al. [40,41] fitted in Reference [31]. All these experimental databases also indicate an increase in U_T by an integral length scale of turbulence.

Curves plotted in dotted-dashed lines in Figure 2 show that a relative increase δA_F

$$\delta A_F = \Lambda^{-2}(c_2 - c_1)^{-1}\langle \iiint |\nabla c|_f(\mathbf{x},t)d\mathbf{x}\rangle \qquad (8)$$

in the area A_F of the reaction-zone surface depends on u'/S_L very similarly to $\langle U_{T,c}\rangle/S_L$. Here, $|\nabla c|_f$ is the value of the Flame Surface Density $|\nabla c|$ conditioned to the reaction zone $c_1 \leq c \leq c_2$, whose boundaries c_1 and c_2 are given by $W(c_1) = W(c_2) = \max\{W(c)\}/2$. In fact, the bending of the $\delta A_F(u'/S_L)$-curves (dotted-dashed lines) is even more pronounced when compared to $\langle U_{T,c}\rangle/S_L$-curves (solid lines), thus, implying a slight increase in $\langle u_c\rangle/S_L$ by u'/S_L. This observation could be attributed to (i) an increase in the reaction rate integrated over the mixing zones ($c \leq c_1$), as such zones are expanded by small-scale turbulent eddies [19], or (ii) an increase in the local reaction rate in the vicinity of cusps [42]. Because the increase in $\langle u_c\rangle/S_L$ by u'/S_L does not contribute to the bending effect, but weakly resists it, further discussion of this trend is beyond the scope of the present communication.

All in all, DNS data reported in Figure 2 indicate that, under conditions of the present study, the bending of $\langle U_{T,c}\rangle/S_L(u'/S_L)$-curves is mainly controlled by the bending of $\delta A_F(u'/S_L)$-curves, in line with the recent finding by Nivarti and Cant [10]. The present DNS data imply that the bending effect stems from the finite thickness of the reaction wave. Indeed, because the sole difference between the previous [23,24] and present simulations consists in substituting the G-equation (3) with Equations (4) and (5), the obtained difference between the linear and bent $\langle U_{T,c}\rangle/S_L(u'/S_L)$-curves should not be attributed to the exploited method of turbulence generation, numerical resolution or scheme, insufficiently large width of the computational domain, etc. The difference between the linear and bent curves can solely be controlled by δ_F. Let us discuss physical mechanisms that could control this difference.

First, the DNSs yield $\langle U_{T,c}\rangle/S_L \approx \delta A_F$, thus, implying that, under conditions of the present study, the influence of turbulence on $\langle U_{T,c}\rangle$ is not controlled by an increase in local burning rate due to enhancement of the local heat and mass transfer by turbulent eddies. It is worth noting, however, that DNS data computed by us at significantly higher $Ka \gg 1$ [26,43] support scaling of $\langle U_{T,c}\rangle/S_L \propto (u'L_{11}/D)^{1/2}$, as predicted by Damköhler [19] by reducing the influence of turbulence on burning rate to the enhancement of the local heat and mass transfer by the turbulence. In other words, the mechanism by Damköhler [19] can play an important role, but under conditions that differ significantly from the conditions of the present study (e.g., the present DNS data show a substantially weaker dependence of $\langle U_{T,c}\rangle$ on L_{11} when compared to the Damköhler's scaling).

Second, eventual reduction in δA_F due to interface-interface or wave-wave collisions [12,13] is taken into account when numerically solving Equation (3) or (4), respectively. Nevertheless, the bending effect is not pronounced in the former case, thus, implying that the collisions do not control the effect under conditions of the present study. If the collisions were of importance, they would yield the bending in the simulations with Equation (3) also.

Third, a physical mechanism that controls the bending effect under conditions of the present study is revealed in Figure 3. In particular, DNS data plotted in lines in Figure 3 indicate that the cumulative probability of finding highly curved reaction zones characterized by a sufficiently large product $|\eta h_m| > b$ of the Kolmogorov length scale and the local iso-surface curvature conditioned to $c_1 \leq c \leq c_2$ (i) is weakly affected by L_{11} if the laminar-wave thickness δ_F is kept constant, cf. dotted-dashed lines which show results obtained in cases A5, B5, and C5, but (ii) is significantly increased when δ_F is decreased, cf. (ii.a) violet dashed line (case TRW) and the three dotted-dashed lines or (ii.b) magenta solid and black double-dashed-dotted lines, which show results computed by solving Equations (3)–(5), respectively, in case C5. Here, b is a positive number of unity order, $h_m = \nabla \cdot \mathbf{n}_G$ and $\mathbf{n}_G = -\nabla G/|\nabla G|$ or $h_m = \nabla \cdot \mathbf{n}_c$ and $\mathbf{n}_c = -\nabla c/|\nabla c|$ in the simulations that deal with Equation (3) or Equations (4) and (5), respectively. Therefore, when the thickness δ_F is decreased, the probability of appearance of small-scale highly curved wrinkles on the reaction-zone surface is increased, thus, increasing the surface area A_F and, hence, the turbulent consumption velocity $\langle U_{T,c}\rangle$. The highest values of A_F and $\langle U_{T,G}\rangle$ are associated with the infinitely thin interface and $\langle U_{T,G}\rangle \propto u'$ in this case, see open symbols and dashed lines in Figure 2. An increase in δ_F results in decreasing $\langle U_{T,c}\rangle$ when compared to $\langle U_{T,G}\rangle \propto u'$, i.e., the bending effect.

DNS data plotted in symbols in Figure 3 support such a scenario by indicating that the cumulative probability of finding highly curved reaction zones characterized by $|\delta_F h_m| > b$ at $c_1 \leq c \leq c_2$ (i) is decreased with increasing L_{11}/δ_F, cf. filled symbols, but (ii) is weakly affected by variations in the velocity field (A or C), Ka, and Kolmogorov scales, provided that u'/S_L, L_{11}/δ_F, and, hence, the Damköhler number are kept constant, cf. red triangles (case A5) and violet crosses (case TRW). Therefore, comparison of lines and symbols in Figure 3 implies that the curvature magnitude is primarily controlled by the thickness δ_F. Because variations in δ_F in cases C5 and TRW result from variations in the diffusivity D, the significant dependence of the curvature magnitude on δ_F implies that it is the molecular transport that impedes wrinkling reaction-zone surface by small-scale eddies in intense turbulence.

In other words, the DNS data plotted in Figure 3 imply that the magnitude of the local curvature of the reaction zone is bounded by the reaction-wave thickness δ_F, whereas smaller-scale turbulent eddies are inefficient in wrinkling the reaction-zone surface. Accordingly, the small-scale range of the turbulence spectrum weakly contributes to an increase in the area A_F of the surface. Such a hypothesis is further supported by approximately equal values of $\langle U_{T,c}\rangle$ obtained in cases A5 and TRW, associated with almost the same Da (which characterizes large-scale eddies), but substantially different Ka (which characterizes the smallest-scale turbulent eddies).

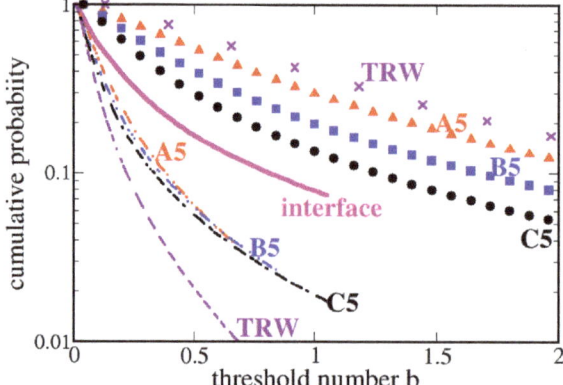

Figure 3. Cumulative probability that the normalized absolute value of local curvature of reaction zone is larger than a positive threshold number b. Lines and symbols show results obtained by normalizing the curvature using the Kolmogorov length scale η and the laminar-wave thickness δ_F, respectively. All results were obtained in cases characterized by the same $u'/S_L = 10$, but different $L_{11}/\delta_F = 2.1$ (cases A5 and TRW), $L_{11}/\delta_F = 3.7$ (case B5), or $L_{11}/\delta_F = 6.7$ (case C5). In case TRW, the laminar reaction-wave thickness δ_F is larger by a factor of four when compared to a reference value used in all other cases associated with Equations (4) and (5). Magenta solid line shows DNS data obtained by solving the level set Equation (3).

To illustrate a mechanism that bounds $|h_m|$ by δ_F, let us follow Zel'dovich et al. [44] and rewrite Equation (4) as follows

$$\partial c/\partial t + u(\partial c/\partial r) - 2(D/r)(\partial c/\partial r) = D(\partial^2 c/\partial r^2) + W \qquad (9)$$

in the spherical coordinate framework. For an expanding reaction wave, the last term on the left-hand side reduces the wave speed when compared to the counterpart planar wave. If the local radius r_w of the wave curvature is on the order of the Kolmogorov length scale η, the magnitude D/r_w of the negative speed resulting from the considered term is on the order of the Kolmogorov velocity (if the Schmidt number Sc = O(1)) and can be much larger than S_L if Ka $\gg$ 1. This curvature-induced speed is negative (positive) for a local wrinkle with the curvature center in products (reactants) and, therefore, tends to damp the local wrinkle. This physical mechanism is controlled by the molecular diffusion and acts even if the local consumption velocity does not depend on the curvature radius r_w. Moreover, while a turbulent eddy whose length scale is significantly smaller than δ_F perturbs the local wave surface during a short lifetime of the eddy, the considered mechanism can smooth the perturbation even after disappearance of the eddy until $D/r_w = O(S_L)$ and $r_w = O(\delta_F)$. Accordingly, if $\eta < \delta_F$, the small-scale range of the entire turbulence spectrum appears to be inefficient in wrinkling the wave surface due to such a smoothing mechanism and this inefficient range expands to smaller length scales when η/δ_F is decreased due to an increase in u'/S_L. This physical mechanism acts to reduce increasing the wave surface area and turbulent consumption velocity with increasing u'/S_L, thus, causing the bending effect.

Indeed, filled black triangles in Figure 4 show that, in case TRW associated with the most pronounced bending effect, the probability of finding high locally negative displacement speed $S_d = (D\nabla^2 c + W)/|\nabla c|$ of the reaction zone is strongly increased by h_m if $|\delta_F h_m| < 1$. The probability of $S_d/S_L < -1$ is larger than 60% if $\delta_F h_m > 2$. Accordingly, if the local curvature of the reaction zone is positive and sufficiently high ($\delta_F h_m > 1$), the zone statistically tends (i) to move to products, i.e., to the curvature center, and, therefore, (ii) to smooth out the local wrinkle of the zone surface. If the local curvature of the reaction zone is negative and sufficiently high ($\delta_F h_m < -1$), the zone moves to reactants (the probability of finding positive $S_d > S_L$ is almost equal to unity, see red stars), i.e., to the curvature center, and, therefore, smooths out the local wrinkle on the zone surface again.

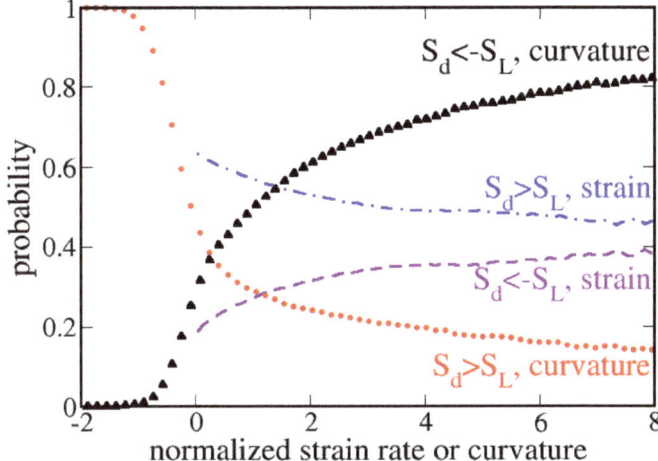

Figure 4. Probabilities of $S_d/S_L < -1$ (violet dashed line and black triangles) and $S_d/S_L > 1$ (blue dotted-dashed line and red circles), conditioned to (i) the local curvature (symbols) h_m of the reaction zone, normalized using δ_F, or (ii) the local total strain (lines) S^2 normalized using $\langle \varepsilon \rangle /(2\nu)$. Case TRW.

On the contrary, violet dashed or blue dotted-dashed line shows that the probability of finding $S_d < -S_L$ or $S_d > S_L$, respectively, in the reaction zone ($c_1 \leq c \leq c_2$) depends weakly on the total strain $S^2 = S_{ij} S_{ij}$. This result implies that turbulent strain rates affect propagation of the reaction zone with respect to the local flow weakly from the statistical viewpoint.

4. Discussion

The above analysis of the present DNS data indicates that the bending of $\langle U_{T,c} \rangle /S_L(u'/S_L)$-curves, computed in the case of $\delta_F > 0$ (see filled symbols in Figure 2) is controlled by the bending of the mean area of the reaction-zone surface as a function of u'/S_L (see dotted-dashed lines in Figure 2). The latter bending is controlled by the following physical mechanism. When a reaction front has a negligible thickness, turbulent eddies of various scales can wrinkle the front surface, increase its area, and, hence, increase turbulent consumption velocity. However, if the local thickness δ_F of the reaction wave is comparable with or larger than the Kolmogorov length scale η, the local molecular transport efficiently smooths out small-scale wrinkles of the reaction-zone surface. Therefore, the local molecular transport impedes increasing the surface area due to the highest local stretch rates created by the smallest turbulent eddies. Consequently, the turbulent consumption velocity U_T is reduced when compared to the case of the infinitely thin interface.

It is worth noting that the above analysis is consistent with DNS data by Wenzel and Peters [45], who simulated self-propagation of a passive interface in constant-density turbulence by solving Equation (3), where S_L was substituted with $S_L - \nu Sc^{-1} \nabla \cdot \mathbf{n}_G$. Indeed, results obtained at $Sc = \infty$, i.e., using the unperturbed laminar-wave speed S_L indicate a linear dependence of δA_F on u' [45], with $U_T = S_L \delta A_F$ being very close to $\langle U_{T,G} \rangle$ obtained by us by solving Equation (3), cf. stars with other symbols in Figure 4 in Reference [24]. However, at $Sc = 0.25$ or 0.50, the bending of $A_F(u')$-curves reported in [45] is well pronounced due to the physical mechanism highlighted above, as the second term in $S_L - \nu Sc^{-1} \nabla \cdot \mathbf{n}_G$ models the smoothing effect of the curvature-induced displacement speed.

The smoothing effect emphasized in the present work differs fundamentally from the well-recognized stretch effect [3–5]. In particular, the smoothing effect (i) is controlled by the molecular diffusivity D of the deficient reactant and (ii) manifests itself in the reduction in A_F independently of a ratio of $\langle u_c \rangle / S_L$. On the contrary, the stretch effect (i) is mainly controlled by the differences in D_F, D_O, and κ in the adiabatic case and (ii) manifests itself in significant variations in the mean local consumption velocity $\langle u_c \rangle$ independently of the area A_F. In the present study, the stretch effect is not pronounced, because the differences in the molecular transport coefficients are not allowed for.

At first glance, the smoothing mechanism emphasized in the present communication appears to be similar to the smoothing effect of flame propagation, which is considered to play an important role by fractal models of premixed turbulent combustion. For instance, the latter mechanism controls the Gibson length scale [46]. However, there are fundamental differences between this well-known propagation-smoothing mechanism and the diffusion-smoothing mechanism emphasized in the present communication. For instance, first, the former and latter mechanisms are associated with the right-hand side and the third term on the left-hand side of Equation (9), respectively. Accordingly, a ratio of magnitudes of effects caused by the two mechanisms may be estimated as follows $r_w S_L / D$. If the radius r_w of the curvature created by the smallest-scale turbulent eddies is significantly less than $\delta_F = D/S_L$, the diffusion-smoothing mechanism should dominate. Second, the propagation-smoothing mechanism can act not only in the case of a reaction wave of a finite thickness, but also in the case of self-propagation of an infinitely thin interface. The fact that the simulations with the G-equation do not yield the bending effect, see open symbols and dashed lines in Figure 2, implies that the propagation-smoothing mechanism is of minor importance under conditions of the present study. It is also worth noting that the present results are qualitatively consistent with experimental data [47] that indicate that the inner cut-off length scale of wrinkles on a flame surface scales as the laminar flame thickness, rather than the Gibson length scale.

In a variable density case, the efficiency of small-scale eddies in wrinkling the reaction zone can be reduced not only due to the smoothing mechanism emphasized in the present communication, but also due to the disappearance of the small-scale eddies due to dilatation and an increase in the mixture viscosity by the temperature. In order to understand what mechanism (smoothing or thermal expansion) dominates and under which conditions, the present DNS study should be extended to flames characterized by substantial density variations.

5. Conclusions

A DNS study of propagation of either an infinitely thin interface or a reaction wave of a nonzero thickness in forced, constant-density, statistically stationary, homogeneous, isotropic turbulence was performed by solving Navier–Stokes equations and either a level set or a reaction-diffusion equation, respectively. In the latter case, the computed mean wave speed $\langle U_{T,c} \rangle$ (i) is reduced when a ratio L_{11}/δ_F of the longitudinal integral length scale L_{11} of the turbulence to the laminar wave thickness δ_F is decreased and (ii) is significantly lower than $\langle U_{T,G} \rangle$ simulated in the former case, with all other things being equal. Moreover, the following results obtained in the present work stem from the finite thickness δ_F of the reaction wave.

First, the computed $\langle U_{T,c} \rangle / S_L (u'/S_L)$-curves show bending. The bending effect is less pronounced at higher L_{11}/δ_F and vanishes in the case of the infinitely thin interface.

Second, under conditions of the present study, the bending effect is controlled by a decrease in the rate of an increase δA_F in the reaction-zone-surface area with increasing u'/S_L. In its turn, the bending of the $\delta A_F(u'/S_L)$-curves stems from inefficiency of small-scale turbulent eddies in wrinkling the reaction-zone surface, because small-scale wrinkles are smoothed out by molecular transport within the local reaction wave. Such a smoothing effect is not pronounced in the case of self-propagation of the infinitely thin interface at a constant speed S_L in statistically the same turbulence.

Author Contributions: Data curation, R.Y.; Formal analysis, R.Y. and A.N.L.; Methodology, R.Y. and A.N.L.; Investigation, R.Y. and A.N.L.; Software, R.Y.; Writing—original draft preparation, A.N.L.; Writing—review and editing, R.Y. and A.N.L.

Funding: The first author (R.Y.) gratefully acknowledges the financial support by the Swedish Research Council (VR) and the National Centre for Combustion Science and Technology (CeCOST). The second author (A.N.L.) gratefully acknowledges the financial support by Chalmers Area of Advance Transport and the Combustion Engine Research Center (CERC).

Acknowledgments: The simulations were performed using the computer facilities provided by the Swedish National Infrastructure for Computing (SNIC) at Beskow-PDC Center. Some results of this work were orally presented at the 10th Mediterranean Combustion Symposium and were reported in the Proceedings of the Symposium, see http://ircserver2.irc.cnr.it/.

Conflicts of Interest: The authors declare no conflict of interest.

References

1. Karpov, V.P.; Sokolik, A.S. Ignition limits in turbulent gas mixtures. *Proc. Acad. Sci. USSR Phys. Chem. Sect.* **1961**, *141*, 866–869.
2. Abdel-Gayed, R.G.; Bradley, D. Dependence of turbulent burning velocity on turbulent Reynolds number and ratio of laminar burning velocity to r.m.s. turbulent velocity. *Symp. Int. Combust.* **1977**, *16*, 1725–1735. [CrossRef]
3. Abdel-Gayed, R.G.; Al-Khishali, K.J.; Bradley, D. Turbulent burning velocities and flame straining in explosions. *Proc. R. Soc. London A* **1984**, *391*, 391–414. [CrossRef]
4. Bray, K.N.C.; Cant, R.S. Some applications of Kolmogorov's turbulence research in the field of combustion. *Proc. R. Soc. London A* **1991**, *434*, 217–240. [CrossRef]
5. Bradley, D. How fast can we burn? *Symp. Int. Combust.* **1992**, *24*, 247–262. [CrossRef]
6. Clavin, P. Dynamical behavior of premixed flame fronts in laminar and turbulent flows. *Prog. Energy Combust. Sci.* **1985**, *11*, 1–59. [CrossRef]
7. Matalon, M. Flame dynamics. *Proc. Combust. Inst.* **2009**, *32*, 57–82. [CrossRef]
8. Klimov, A.M. Laminar flame in a turbulent flow. *ZhPMTF* **1963**, *4*, 49–58.
9. Lipatnikov, A.N.; Chomiak, J. Molecular transport effects on turbulent flame propagation and structure. *Prog. Energy Combust. Sci.* **2005**, *31*, 1–73. [CrossRef]
10. Nivarti, G.V.; Cant, R.S. Direct numerical simulation of the bending effect in turbulent premixed flames. *Proc. Combust. Inst.* **2017**, *36*, 1903–1910. [CrossRef]
11. Zimont, V.L. Theory of turbulent combustion of a homogeneous fuel mixture at high Reynolds number. *Combust. Explos. Shock Waves* **1979**, *15*, 305–311. [CrossRef]
12. Klimov, A.M. Premixed turbulent flames—Interplay of hydrodynamic and chemical phenomena. In *Flames, Lasers and Reactive Systems*; Bowen, J.R., Manson, N., Oppenheim, A.K., Soloukhin, R.I., Eds.; AIAA: New York, NY, USA, 1983; Volume 88, pp. 133–146.
13. Duclos, J.M.; Veynante, D.; Poinsot, T. A comparison of flamelet models for premixed turbulent combustion. *Combust. Flame* **1993**, *95*, 101–117. [CrossRef]
14. Gouldin, F.C. An application of fractals to modelling premixed turbulent flames. *Combust. Flame* **1987**, *68*, 249–266. [CrossRef]
15. Poinsot, T.; Veynante, D.; Candel, S. Quenching processes and premixed turbulent combustion diagrams. *J. Fluid Mech.* **1991**, *228*, 561–606. [CrossRef]
16. Roberts, W.L.; Driscoll, J.F.; Drake, M.C.; Goss, L.P. Images of the quenching of a flame by vortex—To quantify regimes of turbulent combustion. *Combust. Flame* **1993**, *94*, 58–69. [CrossRef]
17. Lipatnikov, A.N.; Chomiak, J. Effects of premixed flames on turbulence and turbulent scalar transport. *Prog. Energy Combust. Sci.* **2010**, *36*, 1–102. [CrossRef]
18. Sabelnikov, V.A.; Lipatnikov, A.N. Recent advances in understanding of thermal expansion effects in premixed turbulent flames. *Annu. Rev. Fluid Mech.* **2017**, *49*, 91–117. [CrossRef]
19. Damköhler, G. Der einfuss der turbulenz auf die ammengeschwindigkeit in gasgemischen. *Z. Electrochem.* **1940**, *46*, 601–652.
20. Venkateswaran, P.; Marshall, A.; Shin, D.H.; Noble, D.; Seitzman, J.; Lieuwen, T. Measurements and analysis of turbulent consumption speeds of H_2/CO mixtures. *Combust. Flame* **2011**, *158*, 1602–1614. [CrossRef]

21. Wabel, T.M.; Skiba, A.W.; Driscoll, J.F. Turbulent burning velocity measurements: Extended to extreme levels of turbulence. *Proc. Combust. Inst.* **2017**, *36*, 1801–1808. [CrossRef]
22. Kerstein, A.R.; Ashurst, W.T.; Williams, F.A. Field equation for interface propagation in an unsteady homogeneous flow field. *Phys. Rev. A* **1988**, *37*, 2728–2731. [CrossRef]
23. Yu, R.; Lipatnikov, A.N.; Bai, X.S. Three-dimensional direct numerical simulation study of conditioned moments associated with front propagation in turbulent flows. *Phys. Fluids* **2014**, *26*, 085104. [CrossRef]
24. Yu, R.; Bai, X.S.; Lipatnikov, A.N. A direct numerical simulation study of interface propagation in homogeneous turbulence. *J. Fluid Mech.* **2015**, *772*, 127–164. [CrossRef]
25. Yu, R.; Lipatnikov, A.N. Direct numerical simulation study of statistically stationary propagation of a reaction wave in homogeneous turbulence. *Phys. Rev. E* **2017**, *95*, 063101. [CrossRef] [PubMed]
26. Yu, R.; Lipatnikov, A.N. DNS study of dependence of bulk consumption velocity in a constant-density reacting flow on turbulence and mixture characteristics. *Phys. Fluids* **2017**, *29*, 065116. [CrossRef]
27. Eswaran, V.; Pope, S.B. An examination of forcing in direct numerical simulations of turbulence. *Comput. Fluids* **1988**, *16*, 257–278. [CrossRef]
28. Lamorgese, A.G.; Caughey, D.A.; Pope, S.B. Direct numerical simulation of homogeneous turbulence with hyperviscosity. *Phys. Fluids* **2005**, *17*, 015106. [CrossRef]
29. Yu, R.; Bai, X.S. A fully divergence-free method for generation of inhomogeneous and anisotropic turbulence with large spatial variation. *J. Comp. Phys.* **2014**, *256*, 234–253. [CrossRef]
30. Bradley, D.; Lau, A.K.C.; Lawes, M. Flame stretch rate as a determinant of turbulent burning velocity. *Phil. Trans. R. Soc. London A* **1992**, *338*, 359–387.
31. Lipatnikov, A.N.; Chomiak, J. Turbulent flame speed and thickness: Phenomenology, evaluation, and application in multi-dimensional simulations. *Prog. Energy Combust. Sci.* **2002**, *28*, 1–74. [CrossRef]
32. Lipatnikov, A.N.; Li, W.Y.; Jiang, L.J.; Shy, S.S. Does density ratio significantly affect turbulent flame speed? *Flow Turbul. Combust.* **2017**, *98*, 1153–1172. [CrossRef] [PubMed]
33. Burluka, A.A.; Griffiths, J.F.; Liu, K.; Orms, M. Experimental studies of the role of chemical kinetics in turbulent flames. *Combust. Explos. Shock Waves* **2009**, *45*, 383–391. [CrossRef]
34. Fogla, N.; Creta, F.; Matalon, M. The turbulent flame speed for low-to-moderate turbulence intensities: Hydrodynamic theory versus experiments. *Combust. Flame* **2017**, *175*, 155–169. [CrossRef]
35. Lipatnikov, A.N.; Chomiak, J.; Sabelnikov, V.A.; Nishiki, S.; Hasegawa, T. A DNS study of the physical mechanisms associated with density ratio influence on turbulent burning velocity in premixed flames. *Combust. Theory Modelling* **2018**, *22*, 131–155. [CrossRef]
36. Smith, K.G.; Gouldin, F.G. Turbulence effects on flame speed and flame structure. *AIAA J.* **1979**, *17*, 1243–1250.
37. Liu, C.C.; Shy, S.S.; Peng, M.W.; Chiu, C.W.; Dong, Y.C. High-pressure burning velocities measurements for centrally-ignited premixed methane/air flames interacting with intense near-isotropic turbulence at constant Reynolds numbers. *Combust. Flame* **2012**, *159*, 2608–2619. [CrossRef]
38. Kido, H.; Kitagawa, T.; Nakashima, K.; Kato, K. An improved model of turbulent mass burning velocity. *Memoirs Faculty Engng, Kyushu University* **1989**, *49*, 229–247.
39. Daniele, S.; Jansohn, P.; Mantzaras, J.; Boulouchos, K. Turbulent flame speed for syngas at gas turbine relevant conditions. *Proc. Combust. Inst.* **2011**, *33*, 2937–2944. [CrossRef]
40. Kobayashi, H.; Tamura, T.; Maruta, K.; Niioka, T.; Williams, F.A. Burning velocity of turbulent premixed flames in a high-pressure environment. *Symp. Int. Combust.* **1996**, *26*, 389–396. [CrossRef]
41. Kobayashi, H.; Kawabata, Y.; Maruta, K. Experimental study on general correlation of turbulent burning velocity at high pressure. *Symp. Int. Combust.* **1998**, *27*, 941–948. [CrossRef]
42. Poludnenko, A.Y.; Oran, E.S. The interaction of high-speed turbulence with flames: Turbulent flame speed. *Combust. Flame* **2011**, *158*, 301–326. [CrossRef]
43. Sabelnikov, V.A.; Yu, R.; Lipatnikov, A.N. Thin reaction zones in highly turbulent medium. *Int. J. Heat Mass Transfer* **2019**, *128*, 1201–1205. [CrossRef]
44. Zel'dovich, Y.B.; Barenblatt, G.I.; Librovich, V.B.; Makhviladze, G.M. *The Mathematical Theory of Combustion and Explosions*; Plenum Publ. Corp.: New York, NY, USA, 1985.
45. Wenzel, H.; Peters, N. Direct numerical simulation and modeling of kinematic restoration, dissipation and gas expansion effects of premixed flames in homogeneous turbulence. *Combust. Sci. Technol.* **2000**, *158*, 273–297. [CrossRef]

46. Peters, N. Laminar flamelet concepts in turbulent combustion. *Symp. Int. Combust.* **1986**, *21*, 1231–1249. [CrossRef]
47. Gülder, Ö.L.; Smallwood, G.J. Inner cutoff scale of flame surface wrinkling in turbulent premixed flames. *Combust. Flame* **1995**, *103*, 107–114. [CrossRef]

© 2019 by the authors. Licensee MDPI, Basel, Switzerland. This article is an open access article distributed under the terms and conditions of the Creative Commons Attribution (CC BY) license (http://creativecommons.org/licenses/by/4.0/).

Article

Effects of Lewis Number on the Evolution of Curvature in Spherically Expanding Turbulent Premixed Flames

Ahmad Alqallaf [1], Markus Klein [2] and Nilanjan Chakraborty [1,*]

[1] School of Engineering, Newcastle University, Newcastle-Upon-Tyne NE1 7RU, UK; A.K.E.R.H.Alqallaf2@newcastle.ac.uk
[2] Department of Aerospace Engineering, Bundeswehr University Munich, LRT1, Werner-Heisenberg-Weg 39, 85577 Neubiberg, Germany; markus.klein@unibw.de
* Correspondence: nilanjan.chakraborty@ncl.ac.uk; Tel.: +44-191-208-3570

Received: 30 November 2018; Accepted: 11 January 2019; Published: 16 January 2019

Abstract: The effects of Lewis number on the physical mechanisms pertinent to the curvature evolution have been investigated using three-dimensional Direct Numerical Simulation (DNS) of spherically expanding turbulent premixed flames with characteristic Lewis number of $Le = 0.8$, 1.0 and 1.2. It has been found that the overall burning rate and the extent of flame wrinkling increase with decreasing Lewis number Le, and this tendency is particularly prevalent for the sub-unity Lewis number (e.g., $Le = 0.8$) case due to the occurrence of the thermo-diffusive instability. Accordingly, the $Le = 0.8$ case has been found to exhibit higher probability of finding saddle topologies with large magnitude negative curvatures in comparison to the corresponding $Le = 1.0$ and 1.2 cases. It has been found that the terms in the curvature transport equation due to normal strain rate gradients and curl of vorticity arising from both fluid flow and flame normal propagation play pivotal roles in the curvature evolution in all cases considered here. The net contribution of the source/sink terms of the curvature transport equation tends to increase the concavity and convexity of the flame surface in the negatively and positively curved locations, respectively for the $Le = 0.8$ case. This along with the occurrence of high and low temperature (and burning rate) values at the positively and negatively curved zones, respectively acts to augment positive and negative curved wrinkles induced by turbulence in the $Le = 0.8$ case, which is indicative of thermo-diffusive instability. By contrast, flame propagation effects tend to weakly promote the concavity of the negatively curved cusps, and act to decrease the convexity of the highly positively curved bulges in the $Le = 1.0$ and 1.2 cases, which are eventually smoothed out due to high and low values of displacement speed S_d at negatively and positively curved locations, respectively. Thus, flame propagation tends to smoothen the flame surface in the $Le = 1.0$ and 1.2 cases.

Keywords: Lewis number; flame curvature; iso-scalar non-material surfaces; turbulent premixed spherical flame

1. Introduction

Spherically expanding turbulent premixed flames are of fundamental importance in Spark Ignition (SI) engines and for understanding accidental explosions. Hence, they are often used as a canonical configuration for laboratory-scale experiments [1–9] and numerical investigations [10–31]. The role of mean flame curvature [20–22,26] on Flame Surface Density (FSD) [20,25] and Scalar Dissipation Rate (SDR) [25,27], which are central to determine the fuel burning rate, has been demonstrated in several previous DNS studies on ignition kernels [11,22] and spherically expanding flames [15,17–27]. Moreover, LES of spherical flames using the flame wrinkling factor [14,16], FSD [27,29] and combined FSD-probability density function (PDF) [28] sub-grid closures, and Reynolds Averaged Navier–Stokes (RANS) simulations using various combustion modelling approaches [10,13,30] showed good agreement with experimental measurements. However, several analyses [20–22,24,27,30,31] demonstrated that there are significant differences between statistically planar and spherical flames, specifically in terms of flame propagation and fuel burning rate. Although these past studies provided important physical insights on spherical flames, they seldom considered thermo-diffusive effects arising from differential diffusion of heat and mass, characterised by Lewis number Le (i.e., ratio of thermal and mass diffusivities). The presence of thermo-diffusive instabilities augments the burning rate as demonstrated in experiments [4,5] and thus spherical flames in lean hydrogen–air mixture grow quicker compared to those in hydrocarbon–air mixture under statistically similar turbulent flow conditions. The necessity to include these effects was demonstrated by computing stoichiometric and fuel-lean hydrogen–air and methane–air spherical flames [30,31].

The flame wrinkling is often characterised in terms of flame front curvature distribution, which plays a key role in determining the local flame propagation behaviour. This is reflected in the correlation between displacement speed S_d and curvature κ_m [21,22,24,32–41]. Moreover, in non-unity Lewis number flames the consumption speed S_c also demonstrates correlation with local flame curvature κ_m [42]. Furthermore, tangential strain rate and curvature have been found to be negatively correlated for small turbulence intensities and the strengths of the correlations of tangential strain rate and flame speed with curvature weaken with increasing turbulence intensity. The flame curvature κ_m is also known to affect the curvature and propagation terms in the Surface Density Function (SDF = $|\nabla c|$ with c being the reaction progress variable) transport equation [43–50], which in turn influence the evolutions of FSD and SDR [27,29–31,50]. It has been demonstrated in several previous analyses [27,39,44,47] that the curvature dependences of displacement speed, temperature and SDF are influenced by the characteristic Lewis number Le, and thus it is expected that the curvature evolution is also likely to be affected by Le.

Pope [51] derived a transport equation for a parameterised surface and recently, Dopazo et al. [52] derived a transport equation of flame front curvature κ_m and analysed the statistical behaviours of the different terms of this transport equation for passive scalar mixing without the effects of heat release. This analysis has been extended by Cifuentes et al. [53] to analyse the statistical behaviours of the terms in the curvature transport equation in a bluff-body stabilised turbulent premixed flame burner configuration using a flame-resolved high-fidelity simulation. However, the effects of characteristic Lewis number on the flame curvature evolution in a configuration with a non-zero mean curvature, as in the case of spherically expanding turbulent premixed flames, are yet to be analysed in detail. For example, the reasons for the differences in flame topology in terms of curvature distribution in response to the variation of Le, and the predominance of positively curved bulges and the presence of intermediate sharply negatively curved cusps between positively curved bulges for the $Le < 1$ flames have not been sufficiently explained in the existing literature. This distribution has strong implications on the augmentation of flame wrinkling and overall burning rate with decreasing Lewis number [39,42,44,54–58]. Thus, it is important to gain a better understanding of the curvature evolution in the presence of thermo-diffusive instabilities and their interrelation with flame shape and propagation in order to be able to derive high fidelity models, which can predict the non-unity Lewis number effects on premixed turbulent combustion. The flame curvature transport equation in Cartesian coordinates is a relatively new concept and the effects of thermo-diffusive instabilities on curvature transport are yet to be understood in detail. In order to address the aforementioned deficits in the existing literature, the statistical behaviours of the curvature transport have been analysed in this paper for statistically spherical flames with characteristic Lewis numbers $Le = 0.8$, 1.0 and 1.2. In this respect, the main objectives of this paper are:

1. To demonstrate the effects of Lewis number on the terms of the transport equation of flame curvature κ_m in spherically expanding turbulent premixed flames.

2. To identify the mechanisms, which lead to the effects of thermo-diffusive instability (i.e., augmentation of burning rate and flame wrinkling and a positive correlation between local burning rate and flame curvature) in the flames with $Le < 1$.

The mathematical and numerical background pertaining to this analysis are presented in the next two sections of this paper. Following that, the results will be presented and subsequently discussed. The main findings will be summarised and conclusions are drawn in the final section of this paper.

2. Mathematical Background

In the current analysis, the chemical mechanism has been simplified by a single-step irreversible reaction so that the effects of characteristic Lewis number Le can be investigated in isolation, as done in several previous analyses [39–42,44,54–58]. The present analysis considers three characteristic Lewis numbers $Le = 0.8$, 1.0 and 1.2 following previous analyses [39–42,44,54–58]. Furthermore, the current analysis only focuses on Lewis number effects (i.e., differential diffusion of heat and mass) and not the differential diffusion between species.

In the case of non-unity Lewis number flames, the scalar field can be characterised by the reaction progress variable c and non-dimensional temperature θ, which are defined as:

$$c = \frac{Y_{R0} - Y_R}{Y_{R0} - Y_{R\infty}} \tag{1}$$

$$\theta = \frac{(T - T_0)}{(T_{ad} - T_0)} \tag{2}$$

where Y_R is the mass fraction of a suitable reactant which is used for defining the reaction progress variable. The subscripts 0 and ∞ are used to refer to values in the unburned gas and fully burned products, respectively. In Equation (2), T, T_0 and T_{ad} denote the dimensional temperature, unburned gas temperature and adiabatic flame temperature, respectively.

The transport equation of the reaction progress variable $c(\mathbf{x}, t)$ is given by:

$$\frac{\partial c}{\partial t} + u_j \frac{\partial c}{\partial x_j} = \frac{1}{\rho}\frac{\partial}{\partial x_j}\left(\rho D_c \frac{\partial c}{\partial x_j}\right) + \dot{w}_c = \frac{1}{\rho}\frac{\partial}{\partial x_N}\left(\rho D_c \frac{\partial c}{\partial x_N}\right) + D_c \frac{\partial c}{\partial x_N} n_{i,i} + \dot{w}_c \tag{3}$$

where u_j is the jth component of the flow velocity, ρ is the fluid density, D_c is the diffusivity of c and $\dot{w}_c$ is chemical reaction rate. The reaction rate $\dot{w}_c$ variation with c for the present thermo-chemistry is shown in Figure 1 of ref. [59] and thus will not be repeated in this paper. The flame normal vector, $\mathbf{n}$, of a c iso-surface is defined as: $n_i = -(\partial c/\partial x_i)/|\nabla c| = -c_{,i}/|\nabla c| = c_{,i}/(\partial c/\partial x_N)$, where x_N is the coordinate in the normal direction to the iso-surface. The quantity, $n_{i,i}/2 = 0.5 \partial n_i/\partial x_i = \kappa_m$ is the mean value of two principal curvatures of the iso-surface and will henceforth be referred to as the flame curvature in this paper. According to the convention used here, the flame normal points towards the reactants and the flame surface has a positive (negative) curvature where it is convex (concave) to the reactants. The transport equation of $c(\mathbf{x}, t)$ can alternatively be presented in the kinematic form as [21,22,24,32–41]:

$$\frac{\partial c}{\partial t} + u_j \frac{\partial c}{\partial x_j} = S_d |\nabla c| \tag{4}$$

where S_d is the displacement speed, which is expressed as [21,22,24,32–41]:

$$S_d(\mathbf{x}, t) = \frac{1}{\rho|\nabla c|}\frac{\partial}{\partial x_N}\left(\rho D_c \frac{\partial c}{\partial x_N}\right) - 2 D_c \kappa_m + \frac{\dot{w}_c}{|\nabla c|} \tag{5}$$

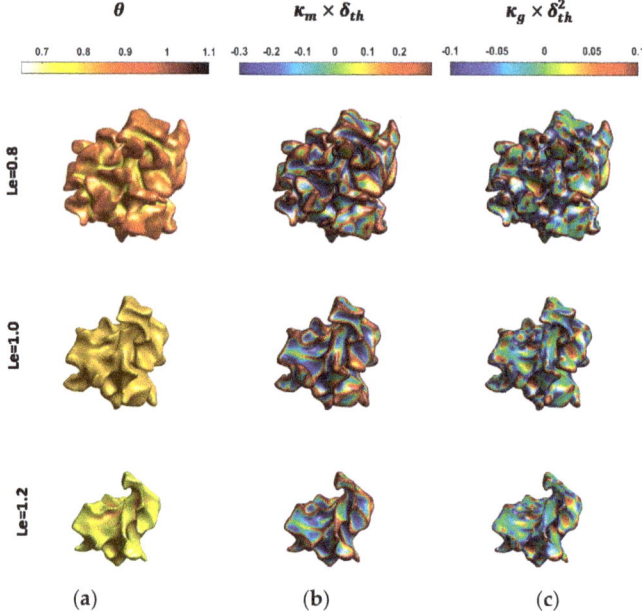

Figure 1. Instantaneous isosurfaces of reaction progress variable $c = 0.8$ colored by non-dimensional temperature θ (**a**), local values of normalised curvature $\kappa_m \times \delta_{th}$ (**b**) and normalised Gauss curvature $\kappa_g \times \delta_{th}^2$ (**c**) for cases with Le = 0.8, 1.0 and 1.2 (1st–3rd row).

Differentiating Equation (3) with respect to x_i yields [52,53]:

$$\frac{1}{|\nabla c|}\left(\frac{\partial |\nabla c|}{\partial t} + v_j^c \frac{\partial |\nabla c|}{\partial x_j}\right) = -n_i v_{j,i}^c n_j \qquad (6)$$

$$\frac{\partial n_i}{\partial t} + v_j^c \frac{\partial n_i}{\partial x_j} = -(\delta_{ij} - n_i n_j) S_{jk}^c n_k + W_{ij}^c n_j; \qquad (7)$$

where $v_j^c = u_j + S_d n_j$ is the jth component of local propagation velocity of a given c isosurface. The quantity $n_i v_{j,i}^c n_j = n_i S_{ij}^c n_j = a_N^c = a_N + \partial S_d / \partial x_N$ is often referred to as the total or effective normal strain rate [46,48,49,53,56,57], whereas $a_N = n_i S_{ij} n_j$ is the fluid-dynamic normal strain rate and $\partial S_d / \partial x_N$ is an added normal strain rate induced by flame propagation [46,48,49,53,60,61]. The gradient of the flame propagation velocity v_j^c takes the following form [52,53]:

$$\begin{aligned} v_{j,i}^c = \frac{\partial v_j^c}{\partial x_i} &= \tfrac{1}{2}\left(\frac{\partial v_j^c}{\partial x_i} + \frac{\partial v_i^c}{\partial x_j}\right) + \tfrac{1}{2}\left(\frac{\partial v_j^c}{\partial x_i} - \frac{\partial v_i^c}{\partial x_j}\right) \\ &= \tfrac{1}{2}\left(\frac{\partial u_j}{\partial x_i} + \frac{\partial u_i}{\partial x_j}\right) + \tfrac{1}{2}\left(\frac{\partial u_j}{\partial x_i} - \frac{\partial u_i}{\partial x_j}\right) + \tfrac{1}{2}\left(\frac{\partial S_d}{\partial x_i} n_j + \frac{\partial S_d}{\partial x_j} n_i\right) \\ &\quad + \tfrac{1}{2}\left(\frac{\partial S_d}{\partial x_i} n_j - \frac{\partial S_d}{\partial x_j} n_i\right) + S_d \tfrac{1}{2}\left(\frac{\partial n_j}{\partial x_i} + \frac{\partial n_i}{\partial x_j}\right) + S_d \tfrac{1}{2}\left(\frac{\partial n_j}{\partial x_i} - \frac{\partial n_i}{\partial x_j}\right) \\ &= S_{ij}^c - W_{ij}^c \end{aligned} \qquad (8)$$

Table 1 presents the nomenclature associated with Equation (8), where the total strain rate and rotation rate tensors are expressed as $S_{ij}^c = S_{ij} + S_{ij}^a$ and $W_{ij}^c = W_{ij} + W_{ij}^a$ respectively [52,53]. The strain rate S_{ij} and rotation rate W_{ij} tensors originate due to the fluid motion, and the additional strain rate S_{ij}^a and rotation W_{ij}^a rate tensors originate due to the gradients of spatially dependent S_d.

Table 1. Nomenclature associated with the velocity gradient tensor $v_{i,j}^c$ of an iso-surface element.

Description	Term
Total strain rate tensor	$S_{ij}^c = 0.5\left(\partial v_i^c/\partial x_j + \partial v_j^c/\partial x_i\right)$
Total rotation rate tensor	$W_{ij}^c = 0.5\left(\partial v_i^c/\partial x_j - \partial v_j^c/\partial x_i\right)$
Flow strain rate tensor	$S_{ij} = 0.5\left(\partial u_i/\partial x_j + \partial u_j/\partial x_i\right)$
Flow rotation rate tensor	$W_{ij} = 0.5\left(\partial u_i/\partial x_j - \partial u_j/\partial x_i\right)$
Added strain rate tensor	$S_{ij}^a = \underbrace{0.5\left[\left(\partial S_d/\partial x_j\right)n_i + (\partial S_d/\partial x_i)n_j\right]}_{\text{Space dependence of } S_d} + \underbrace{S_d 0.5\left(\partial n_i/\partial x_j + \partial n_j/\partial x_i\right)}_{\text{Propagating curved iso-}c}$
Added rotation rate tensor	$W_{ij}^a = \underbrace{0.5\left[\left(\partial S_d/\partial x_j\right)n_i - (\partial S_d/\partial x_i)n_j\right]}_{\text{Space dependence of } S_d} + \underbrace{S_d 0.5\left(\partial n_i/\partial x_j - \partial n_j/\partial x_i\right)}_{\text{Propagating curved iso-}c}$

Taking the derivative with respect to x_i on both sides of Equation (7) yields a transport equation of $\kappa_m = 0.5(\kappa_1 + \kappa_2) = 0.5 n_{i,i}$ (with κ_1 and κ_2 being the two principal curvatures) [52,53]:

$$\frac{\partial \kappa_m}{\partial t} + v_j^c \frac{\partial \kappa_m}{\partial x_j} = \underbrace{\frac{a_N(n_{i,i})}{2}}_{T_1} + \underbrace{\frac{1}{2}\frac{\partial a_N}{\partial x_N}}_{T_2} \underbrace{-S_{ij}n_{j,i}}_{T_3} \underbrace{-\frac{1}{2}\frac{\partial S_{ij}}{\partial x_i}n_j}_{T_4} + \underbrace{\frac{1}{2}\frac{\partial W_{ij}}{\partial x_i}n_j}_{T_5}$$
$$\underbrace{}_{\text{flow terms}}$$
$$+ \underbrace{\frac{1}{2}\frac{\partial S_d}{\partial x_N}n_{i,i}}_{T_6} + \underbrace{\frac{1}{2}\frac{\partial^2 S_d}{\partial x_N^2}}_{T_7} \underbrace{-S_{ij}^a n_{j,i}}_{T_8} \underbrace{-\frac{1}{2}\frac{\partial S_{ij}^a}{\partial x_i}n_j}_{T_9} + \underbrace{\frac{1}{2}\frac{\partial W_{ij}^a}{\partial x_i}n_j}_{T_{10}}$$
$$\underbrace{}_{\text{added terms}}$$
(9)

The terms T_{1-5} arise due to fluid motion, whereas T_{6-10} originate due to flame propagation. The positive (negative) contributions of these terms tend to increase the convexity (concavity) of iso-surfaces. The physical significances of the terms on the right side of Equation (9) are summarised in Table 2.

Table 2. Description of the various terms in the curvature transport Equation (9).

Flow Terms		Added Terms	
Terms	Description	Terms	Description
T_1	Contribution due to curvature and flow normal strain rate correlation	T_6	Term due to curvature and added normal strain rate correlation
T_2	Contribution due to flow normal strain rate normal variation	T_7	Contribution due to added normal strain rate normal variation
T_3	Flow stretching term	T_8	Added stretching term
T_4	Contribution of flow strain rate gradients	T_9	Contribution of added strain rate gradients
T_5	Contribution of flow vorticity curl	T_{10}	Contribution of added vorticity curl

It can be appreciated from Equation (9) that curvature transport depends mainly on the statistics of fluid velocity/vorticity, scalar gradient and displacement speed. It has been demonstrated in the past that displacement speed statistics from simple chemistry [38–41,45] and detailed chemistry [32–37] DNS are qualitatively similar. The same is true for the statistics of the reactive scalar gradient obtained from simple chemistry [43–45,62] and detailed chemistry [47,62] DNS studies. Moreover, the vorticity and sub-grid flux statistics obtained from simple chemistry [59,63,64] DNS are found to be qualitatively consistent with those obtained from detailed chemistry [65,66] DNS. Furthermore, several models developed based on simple chemistry data [64,67,68] have been found to perform equally well in the context of detailed chemistry and transport [66,69,70].

3. Numerical Implementation

In the present analysis, DNS simulations of spherically expanding turbulent premixed flames with $Le = 0.8$, 1.0 and 1.2 have been carried out using a well-known compressible code SENGA [71] where the conservation equations of mass, momentum, energy and reaction progress variable have been solved in non-dimensional form. The $Le = 0.8$ case is representative of hydrogen-blended methane-air mixtures (e.g., 10% by volume hydrogen blended methane-air flames with overall equivalence ratio of 0.6) and the Lewis number 1.2 case is representative of a hydrocarbon-air mixture involving a hydrocarbon fuel which is heavier than methane (e.g., ethylene-air mixture with equivalence ratio of 0.7) [72,73]. The unity Lewis number flames are analogous to the stoichiometric methane-air flame [72,73]. The spatial discretisation for internal grid points has been carried out using a 10th order central difference scheme and the order of differentiation decreases gradually to a one-sided 2nd order scheme at the non-periodic boundaries [71]. The time-advancement has been carried out using a 3rd order explicit Runge–Kutta scheme [74]. The computational domain is taken to be a cube of $58.10\delta_{th} \times 58.10\delta_{th} \times 58.10\delta_{th}$ (where $\delta_{th} = (T_{ad} - T_0)/\max|\nabla T|_L$ is the thermal flame thickness and the subscript 'L' is used to refer to conditions in the unstrained laminar premixed flame), which is discretised by a uniform Cartesian grid of $650 \times 650 \times 650$. In all the cases considered here, the boundaries of the computational domain are taken to be partially non-reflecting and are specified using the Navier–Stokes Characteristic Boundary conditions (NSCBC) technique [75]. The reacting scalar fields obtained from a steady state unstrained laminar flame have been utilised to create a burned gas sphere with its centre initially at the centre of the domain. This reacting scalar field is allowed to evolve in a quiescent environment at least for one chemical time scale (i.e., $t = \delta_{th}/S_L$). For all simulations, standard values are taken for Prandtl number Pr and Zel'dovich number $\beta = T_{ac}(T_{ad} - T_0)/T_{ad}^2$ (i.e., $Pr = 0.7$ and $\beta = 6.0$ with T_{ac} being the activation temperature) and the heat release parameter $\tau = (T_{ad} - T_0)/T_0$ is taken to be 4.5 for all cases considered here. The spherical laminar flame kernels for different Lewis numbers with a normalised radius of $rS_L/\alpha_{T0} = 10.6$ (where α_{T0} is the thermal diffusivity of the unburned gas) based on the region corresponding to $c \geq 0.85$ have been considered as the initial condition for turbulent simulations. A standard pseudo-spectral method [76] has been adopted to initialise homogeneous isotropic turbulent velocity fluctuations following the model spectrum by Pope [77]. For all cases the initial values of the normalised root-mean-square velocity fluctuation and longitudinal integral length scale are given by: $u'/S_L = 7.5$ and $l/\delta_{th} = 4.58$, respectively. These values of u'/S_L and l/δ_{th} are representative of the thin reaction zones regime of combustion according to the regime diagram by Peters [78]. All the turbulent simulations have been continued for 2 initial eddy turnover times (i.e., $t = l/\sqrt{k}$ where k is the turbulent kinetic energy based on the whole domain), which is equivalent to 1.0 chemical timescale (i.e., $t_{chem} = \delta_{th}/S_L$). By this time, the turbulent kinetic energy was not varying rapidly with time and u' decayed by 40% in comparison to its initial value.

4. Results and Discussion

4.1. Curvature Characterisation

The flame topology is often characterised by the mean of principal curvatures (i.e., $\kappa_m = 0.5(\kappa_1 + \kappa_2)$) and Gauss curvature $\kappa_g = \kappa_1\kappa_2$ [79] where κ_1 and κ_2 are the principal curvatures. The $c = 0.8$ isosurface coloured by local values of non-dimensional temperature θ, normalised mean curvature $\kappa_m \times \delta_{th}$ and normalised Gauss curvature $\kappa_g \times \delta_{th}^2$ for the $Le = 0.8$, 1.0 and 1.2 flames are shown in Figure 1. It is worth noting that the reaction rate $\dot{w}_c$ assumes maximum value close to $c = 0.8$ for the present thermochemistry and this isosurface can be taken to be the flame surface for the purpose of this analysis.

It is evident from Figure 1 that the extent of wrinkling and burned gas volume increase with decreasing Le, and especially the occurrences of saddle point topologies (i.e., $\kappa_g < 0$) and sharply negatively curved cusps (i.e., large magnitudes of negative κ_m) increase with decreasing Lewis number.

The augmentation of flame wrinkling with decreasing Le has implications on the extent of burning and flame surface area, which can be substantiated from Table 3 where the normalised volume-integrated burning rate Ω_T/Ω_L and flame surface area A_T/A_L are listed. Here, Ω and A are evaluated by volume-integrals $\Omega = \int_V \rho \dot{w}_c dV$ and $A = \int_V |\nabla c| dV$, respectively and the subscripts T and L refer to the values in turbulent flame and initial laminar flame kernels, respectively. The fresh reactants diffuse faster into the reaction zone than the rate at which heat diffuses out in the $Le = 0.8$ flame. This gives rise to higher burning rate in the $Le = 0.8$ flame than in the corresponding unity Lewis number case. The diffusion of reactants is slower into the reaction zone than the rate at which heat diffuses out in the $Le = 1.2$ flame, which in turn reduces the burning rate in the $Le = 1.2$ case than in the corresponding unity Lewis number case. This is qualitatively consistent with several previous findings [1,39,41,42,44,54–58].

Table 3. Lewis number dependence of the normalised volume-integrated burning rate Ω_T/Ω_L and flame surface area A_T/A_L when the statistics were extracted.

Le	Ω_T/Ω_L	A_T/A_L
0.8	13.75	13.22
1.0	6.86	7.30
1.2	3.86	4.66

It can be seen from Figure 1 that high temperature zones are associated with positive κ_m values in the $Le = 0.8$ case, whereas the high temperature zones are associated with negatively curved zones (i.e., $\kappa_m < 0$) in the $Le = 1.2$ case. The non-dimensional temperature θ remains uniform and equal to 0.8 (i.e., $\theta = c = 0.8$) on the $c = 0.8$ isosurface in the unity Lewis number case. The statistically spherical flames have predominantly positive curvatures where the combination of strong focussing of reactants and weak defocussing of heat gives rise to high reaction rate in the positively curved regions in the $Le = 0.8$ flame. Just the opposite mechanisms lead to low reaction rate at the negatively curved regions in the $Le = 0.8$ flame. This also gives rise to $\theta > c$ at the positively curved locations in the $Le = 0.8$ flame. As a result of this, the $Le = 0.8$ case exhibits stable positively curved bulges with large radius of curvature separated by sharply negatively curved cusps. In the $Le = 1.2$ case, the focussing of heat is faster than the defocussing of reactants at the negatively curved locations, which increases the reaction rate magnitude in these locations and may give rise to local occurrences of $\theta > c$. Thus, the sharply negatively curved pockets burn faster in the $Le = 1.2$ case and as a result sharp negatively curved cusps with small radius of curvature are unlikely to survive in this case. The unity Lewis number case is thermo-diffusively neutral and thus, for low-Mach number adiabatic conditions, the non-dimensional temperature θ remains equal to the reaction progress variable.

The high temperature values at the positively curved zones significantly increase the fuel consumption rate per unit area in the turbulent $Le = 0.8$ case in comparison to the corresponding unstrained laminar flame value and this gives rise to a Ω_T/Ω_L value, which is greater than A_T/A_L. The consumption rate per unit area in the turbulent $Le = 1.0$ case remains comparable to the corresponding unstrained laminar flame value and thus Ω_T/Ω_L and A_T/A_L remain close to each other. The combination of strong thermal diffusion and weak diffusion of reactants into the reaction zone reduces the consumption rate per unit area in the positively stretched zones in turbulent $Le = 1.2$ case in comparison to the corresponding unstrained laminar flame value, which results in $\Omega_T/\Omega_L < A_T/A_L$ in this case.

The probability density functions (PDFs) of normalised curvature $\kappa_m \times \delta_{th}$ for different c-isosurfaces across the flame front are shown in Figure 2. It is evident from Figure 2 that the width of the curvature PDFs tends to increase with decreasing Lewis number due to increased flame wrinkling with decreasing Lewis number Le. The most probable value of κ_m has been found to be close to zero for all flame kernels considered here but the mean value of κ_m remains positive due to the statistically spherical configuration. In the $Le \geq 1$ flames, the PDFs of normalised curvature $\kappa_m \times \delta_{th}$ remain almost symmetric around $\kappa_m \times \delta_{th} = 0$ towards the unburned gas side of the flame front (e.g., $c = 0.1$) due to the strong deformation of the flame surface as a result of the interaction between energetic eddies with the preheat zone. However, the PDFs of normalised curvature $\kappa_m \times \delta_{th}$ exhibit higher propensity to obtain positive values for the major part of flame front. It can be seen from Figure 2 that the high magnitudes of negative κ_m are more frequent for the $c = 0.5$, 0.7 and 0.9 isosurfaces in the $Le = 0.8$ case than in the corresponding $Le = 1.0$ and 1.2 cases. This tendency for the $Le = 0.8$ flame could be a consequence of thermo-diffusive instability associated with $Le < 1$, where sharply negative curved cusps with small radii of curvature are found in between positively curved bulges with large or moderate radii of curvature.

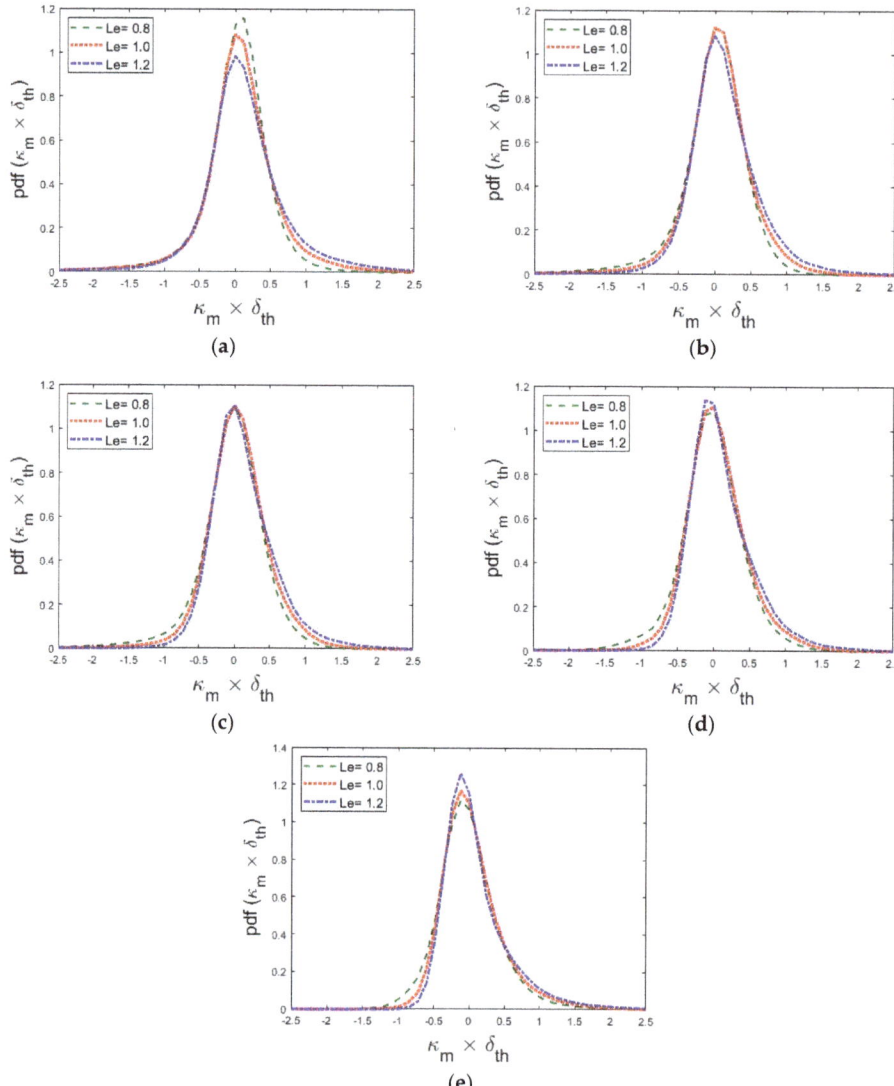

Figure 2. Probability density functions (PDFs) of $\kappa_m \times \delta_{th}$ on (**a**–**e**) $c = 0.1$, 0.3, 0.5, 0.7 and 0.9 isosurfaces for Le = 0.8, 1.0 and 1.2.

The Joint Probability Density Functions (JPDFs) of normalised mean curvature $\kappa_m \times \delta_{th}$ and normalised Gauss curvature $\kappa_g \times \delta_{th}^2$ for different c iso-surfaces across the flame are shown in Figure 3. The region corresponding to $\kappa_g > \kappa_m^2$ on the $\kappa_m - \kappa_g$ plane corresponds to complex principal curvature values and thus cannot be realised in practice. The combination of $\kappa_m > 0$ ($\kappa_m < 0$) and $\kappa_g > 0$ corresponds to cup convex (concave) shapes. By contrast, the combination of $\kappa_m > 0$ ($\kappa_m < 0$) and $\kappa_g < 0$ corresponds to saddle convex (concave) shapes. Finally, the combination of $\kappa_m > 0$ ($\kappa_m < 0$) and $\kappa_g = 0$ corresponds to tile convex (concave) flame topologies. For a schematic diagram of these topologies interested readers are referred to Figure 1 of ref. [53]. The spread of $\kappa_m \times \delta_{th}$ values on both positive and negative sides for the $Le = 0.8$ case is greater than that in the corresponding $Le = 1.0$ and 1.2 cases, which is indicative of larger extent of flame wrinkling in the $Le = 0.8$ case (see Figure 1 and Table 3). The highest values of joint PDFs are obtained for a positive value of κ_m close to $\kappa_m \approx 0$, which is consistent with the observations made from Figure 2. Figure 3 demonstrates that the joint PDFs in the $Le = 0.8$ case show skewness for the negative values of $\kappa_m \times \delta_{th}$ on the unburned gas side of the flame, but the joint PDFs eventually become skewed towards the positive values of κ_m as the burned gas side is approached. The distribution depicted by the joint PDF also shows a propensity to obtain predominantly positive κ_m towards the burned gas side of the flame front for both $Le = 1.0$ and 1.2 cases but this tendency is more prevalent for the $Le = 1.2$ case. The high probability of finding negatively curved cusps ($\kappa_m < 0$) for the $Le = 0.8$ case is a reflection of the thermo-diffusive instability in this flame. Both cup and saddle topologies appear considerably for convex and concave curvatures throughout the flame front in all cases. However, the cup like topology, whether for concave or convex structure, is more probable than the saddle and tile topologies for all cases considered here. The probability of finding saddle topologies decreases from the unburned to the burned gas side for all flames considered here. Moreover, the probability of finding saddle topologies decreases with increasing Le, which is consistent with the observations made from Figure 1.

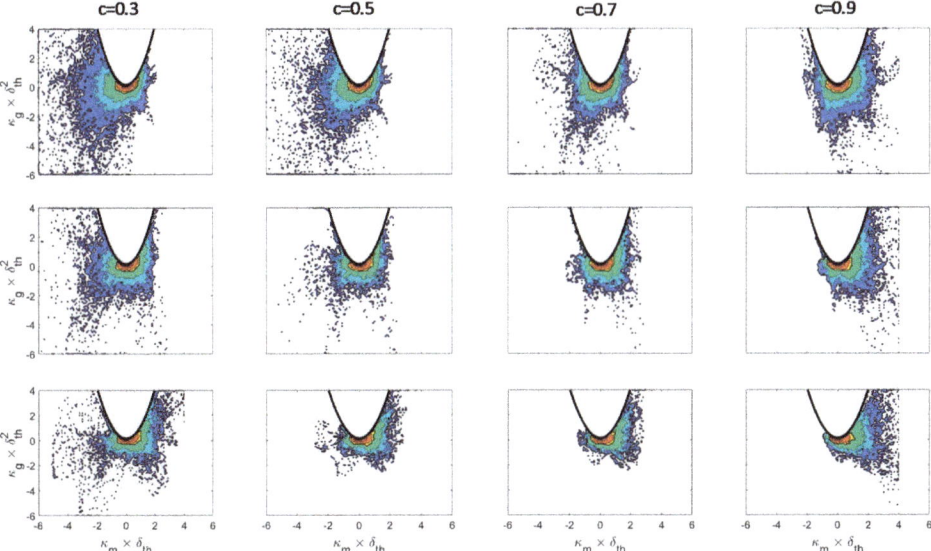

Figure 3. Joint PDF (coloured from blue to dark red) between normalised curvature $\kappa_m \times \delta_{th}$ and normalised Gauss curvature $\kappa_g \times \delta_{th}^2$ for $c = 0.3, 0.5, 0.7$ and 0.9 (1st–4th column) isosurfaces for cases $Le = 0.8, 1.0$ and 1.2 (1st–3rd row).

The budgets of the various terms in the curvature transport equation conditional on c will be discussed in the next section. Based on this analysis it will be possible to judge the magnitude of the different terms and which terms possibly balance each other. Beside this, another focus of this paper is to establish a relation between the terms of the curvature transport equation and flame instabilities that might occur under certain conditions. It will therefore be instructive to analyse the terms $T_1, \ldots, T_{10}$ conditional on mean curvature in the next but one section. Only by this additional representation, it will be possible to understand if positive or negative flame curvature elements are damped or possibly amplified in the presence of flame instabilities.

4.2. Mean Profiles of Source/Sink Terms of the Curvature Transport Equation

The profiles of mean values of $T_i \times \delta_{th}^2/S_L$ conditional on c for $Le = 0.8$, 1.0 and 1.2 cases are shown in Figure 4a–c respectively. The objective of Figure 4 is to provide insights into the mean behaviour of the terms on the right-hand side of the curvature transport equation (i.e., Equation (9)) across the flame. It can be seen from Figure 4 that the term $T_1 \times (\delta_{th}^2/S_L)$, originating from the correlation between curvature and normal strain rate, assumes weakly negative values for the major part of the flame front for the for $Le = 1.0$ and 1.2 cases. In these cases, the mean value of $T_1 \times (\delta_{th}^2/S_L)$ remains weakly positive towards the burned gas side of the flame front. However, the mean values of $T_1 \times (\delta_{th}^2/S_L)$ remain positive throughout the flame front for the $Le = 0.8$ case. Therefore, the term T_1 tends to increase the convexity of iso-surface in the $Le = 0.8$ case, while the opposite behaviour is obtained in the $Le = 1.0$ and 1.2 cases. In the $Le = 1.0$ and 1.2 flames the reaction progress variable gradient ∇c aligns predominantly with the most compressive principal strain rate due to stronger turbulent straining than the strain rate arising from flame normal acceleration induced by chemical heat release. This leads to predominantly negative normal strain rate $a_N = (e_\alpha \cos^2 \theta_\alpha + e_\beta \cos^2 \theta_\beta + e_\gamma \cos^2 \theta_\gamma)$ (where e_α, e_β and e_γ are the most extensive, intermediate and the most compressive eigenvalues of the strain rate tensor $S_{ij} = 0.5(\partial u_i/\partial x_j + \partial u_j/\partial x_i)$ and $\theta_\alpha, \theta_\beta$ and θ_γ are the angles of their eigenvectors with ∇c, respectively) in the $Le = 1.0$ and 1.2 flames considered here. In these cases, ∇c exhibits some tendency for local preferential collinear alignment with e_α only in the heat release zone. However, the alignment of ∇c with the most extensive principal strain rate weakens especially in the positively curved locations due to defocussing of heat [27,39,40,43,80–82], which gives rise to a situation where the highly negative a_N values are associated with highly positively curved locations. This gives rise to negative mean value of $T_1 \times (\delta_{th}^2/S_L)$ in the $Le = 1.0$ and 1.2 cases considered here. In the case of $Le = 0.8$, the alignment of ∇c with the eigenvector associated with e_α (e_γ) is stronger (weaker) than the $Le = 1.0$ and 1.2 cases owing to stronger heat release effects. Furthermore, the alignment of ∇c with the eigenvector associated with e_α increases in the positively curved regions due to high reaction rate and heat release associated with these locations. As a result, high positive values of a_N are obtained at the positively curved locations in the $Le = 0.8$ case to yield positive mean values of $T_1 \times (\delta_{th}^2/S_L)$.

The mean contribution to the curvature transport due to normal gradients of the flow normal strain rate, $T_2 \times (\delta_{th}^2/S_L) = [(\partial a_N/\partial x_N)/2]/(\delta_{th}^2/S_L)$, assumes negative values on the reactant side but becomes positive towards the burned gas side of the flame. The normal strain rate a_N increases gradually from the reactant side toward the product side due to the heat release and reaches its maximum close to the reaction zone, and hence its normal gradient $\partial a_N/\partial x_N$ becomes zero, then it decreases again near the product side [61]. Accordingly, the tendency of $(\partial a_N/\partial x_N) < 0$ is high on the reactant side, while the opposite occurs on the product side of the flame front. Consequently, T_2 tends to act as a sink (source) term toward the reactant (product) side of the flame front. However, the profile and the magnitude of a_N change due to differential diffusion effects induced by non-unity Lewis number [61]. This also alters the location at which the highest value of a_N is obtained [61]. As a result of this, the magnitude of T_2 increases with decreasing Le and in all cases the magnitude of the positive mean value of $T_2 \times (\delta_{th}^2/S_L)$ remains greater than the negative mean contribution towards the unburned gas side.

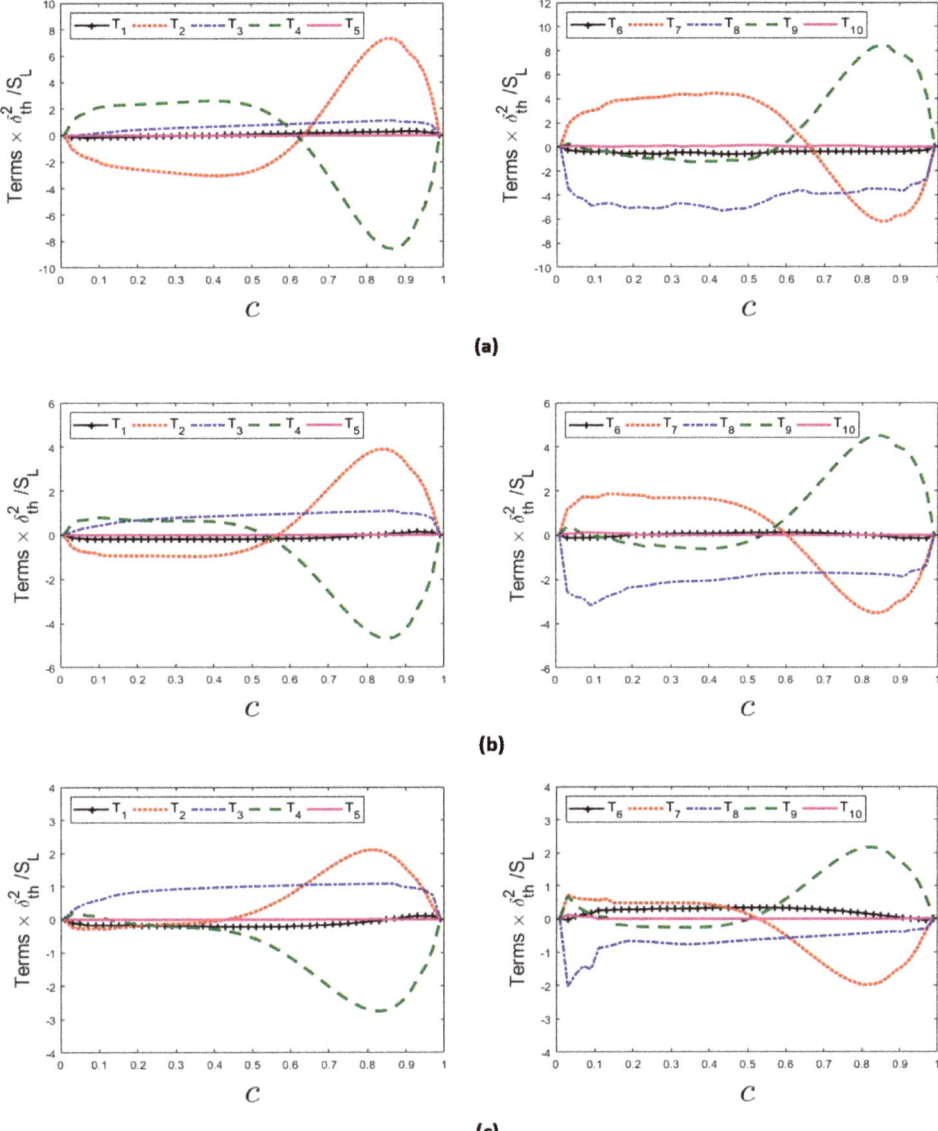

Figure 4. Profiles of the normalised mean values of the fluid flow induced terms (i.e., $T_{1-5} \times \delta_{th}^2/S_L$) (left column) and flame propagation induced terms (i.e., $T_{6-10} \times \delta_{th}^2/S_L$) (right column) of the curvature transport equation conditional upon c for cases (**a**–**c**) Le = 0.8, 1.0 and 1.2.

The mean value of the curvature flow stretching term $T_3 \times (\delta_{th}^2/S_L) = -(S_{ij}n_{j,i})(\delta_{th}^2/S_L)$ is found to be positive throughout the flame brush in all cases considered here. The magnitude of T_3 becomes comparable to T_2 for the Le = 1.2 case but its magnitude in comparison to T_2 remains small in the Le = 0.8 case. Using (x_{T1}, x_{T2}) as the local principal axes along the tangential directions on a given c isosurface, the curvature flow stretching term can be expressed as $T_3 = -(S_{11}\kappa_1 + S_{22}\kappa_2)$, where S_{11} and S_{22} are the tangential strain rates along axes x_{T1} and x_{T2}, respectively. The mean tangential strain rate $a_T = S_{11} + S_{22}$ assumes positive values throughout the flame front for all flames considered here [43–49]. A positive magnitude of T_3 is obtained for a combination of cup and saddle concave iso-scalar topologies. For $S_{11} > 0$ and $S_{22} > 0$, one can obtain $T_3 > 0$ for a cup concave structure where $\kappa_1 < 0$ and $\kappa_2 < 0$ or for a saddle topology where either $\kappa_1 < 0$ and $\kappa_2 > 0$ when $S_{11} > S_{22}$ or $\kappa_1 > 0$ and $\kappa_2 < 0$ when $S_{11} < S_{22}$.

The statistical behaviour of the mean value of the term $T_4 \times (\delta_{th}^2/S_L) = -\{[(\partial S_{ij}/\partial x_i)n_j]/2\}(\delta_{th}^2/S_L)$ conditional on c shows an opposite trend in comparison to that of T_2, in other words it exhibits positive contribution toward the reactant side but becomes negative on the burned gas side. This means that the two vectors with components $\partial S_{ij}/\partial x_i$ and n_j, respectively are in the same direction on the reactant side, while they point in the opposite direction on the burned gas side. The magnitudes of T_2 and T_4 are mostly comparable and remain in balance for all cases considered here.

The normalised contribution to the curvature transport due to vorticity gradients, $T_5 \times (\delta_{th}^2/S_L) = \{[(\partial W_{ij}/\partial x_i)n_j]/2\}(\delta_{th}^2/S_L)$, can alternatively be written as: $T_5 = \{[n_i \varepsilon_{ijk}(\partial \omega_k/\partial x_j)]/4\}(\delta_{th}^2/S_L)$. The mean value of T_5 can be expressed as $T_5 = (\partial \omega_2/\partial x_1 - \partial \omega_1/\partial x_2)/4$, using the local principal axes (x_{T1}, x_{T2}, x_N) and $\mathbf{n} = (0,0,1)$, where ω_1 and ω_2 are the components of the flow vorticity tangent to the c iso-surface. Thus, co-rotating parallel vortices of different intensity and counter-rotating parallel vortices of the same intensity can potentially curve a planar local iso-surface structure, leading to positive or negative curvatures. For the cases considered here, the mean value of T_5 shows negligible contribution throughout the flame front.

In the $Le = 1.2$ case, the mean values of the term $T_6 \times (\delta_{th}^2/S_L) = \{(\partial S_d/\partial x_N)n_{i,i}/2\}(\delta_{th}^2/S_L)$ which arises due to the correlation between $(\partial S_d/\partial x_N)$ and κ_m, exhibit positive contributions throughout the flame front, while a weakly positive trend is seen in the $Le = 1.0$ case for the major part of the flame. However, this term shows negative contribution throughout the flame in the $Le = 0.8$ case. It has been demonstrated elsewhere that the contributions of $(\partial S_d/\partial x_N)$ and a_N are of same order of magnitude but opposite in behaviour [61], and hence an opposite behaviour for the mean values of T_1 and T_6 is observed. It is also worth noting that $(\partial S_d/\partial x_N)$ is predominantly negative and its magnitude decreases with increasing Le [61]. Furthermore, small values of the flame thickness are associated with the positively curved locations in the $Le = 0.8$ case [60], which tends to increase the magnitude of negative $(\partial S_d/\partial x_N)$ at positive κ_m values. This leads to predominantly negative mean contribution of T_6 in the $Le = 0.8$ flame. By contrast, small values of the flame thickness are associated with the negatively curved locations in the $Le = 1.2$ case [61], and this increases the magnitude of negative $(\partial S_d/\partial x_N)$ at negative κ_m values, leading to positive mean values of T_6. The quantity $(\partial S_d/\partial x_N)$ remains weakly correlated with κ_m in the $Le = 1.0$ case [61], which leads to weak positive mean value of this term throughout the flame front.

The normalised mean value of the term $T_7 \times (\delta_{th}^2/S_L) = [(\partial^2 S_d/\partial x_N^2)/2](\delta_{th}^2/S_L)$, due to the normal gradient of the added normal strain rate, assumes a positive mean value towards the unburned gas side but becomes negative toward the burned gas side of the flame. The magnitude of T_7 decreases with increasing Le. As it has been shown elsewhere [61] that the contributions of $(\partial S_d/\partial x_N)$ and a_N are of the same order of magnitude but opposite in sign, their normal gradients $(\partial^2 S_d/\partial x_N^2$ and $\partial a_N/\partial x_N)$ are expected to exhibit the opposite behaviour, and thus, T_7 exhibits an opposite trend to T_2.

In the thin reaction zones regime flames considered here, the normalised added stretching term $T_8 \times (\delta_{th}^2/S_L) = -(S_{ij}^a n_{j,i})(\delta_{th}^2/S_L)$ acts as a leading order contributor in all cases where it remains negative throughout the flame. Using the local principal axes (x_{T1}, x_{T2}, x_N) and for $\mathbf{n} = (0,0,1)$, one can express the term as $T_8 = -S_d(\kappa_1^2 + \kappa_2^2)$, which explains the predominant negative contribution of T_8 throughout the flame front. The magnitude of T_8 increases with decreasing Le due to a larger spread of displacement speed S_d and principal curvatures κ_1 and κ_2 as a result of greater extent of flame wrinkling.

The second derivatives of S_d and n_i determine the normalised mean contributions due to the added strain rate gradients $T_9 \times (\delta_{th}^2/S_L) = -\{[(\partial S_{ij}^a/\partial x_i)n_j]/2\}(\delta_{th}^2/S_L)$, which exhibit negative (positive) contribution towards reactant (product) side in all cases. The mean contribution of the added vorticity contribution term $T_{10} = [-0.25(\delta_{ij} - n_i n_j)\partial^2 S_d/\partial x_i \partial x_j + 0.25(\partial S_d/\partial x_N)n_{i,i}]$ remains negligible in all cases considered here.

It can be inferred from Figure 4 that the terms T_2, T_4, T_7, T_8 and T_9 are the leading order terms of the mean curvature transport for all cases. In addition, the term T_3 plays a leading order role for the $Le = 1.2$ case. However, in all cases considered here, the mean contributions of T_2 and T_4 assume comparable magnitudes with opposite signs and remain mostly in balance.

4.3. Mean Profiles of the Terms of the Curvature Transport Equation Conditioned Upon Curvature

The normalised mean values of flow-induced terms (i.e., T_{1-5}) and added flame propagation terms (i.e., T_{6-10}) in the mean curvature transport equation conditional upon the normalised curvature $\kappa_m \delta_{th}$ for the $c = 0.8$ isosurface are shown in Figure 5. It can be seen from Figure 5 that the magnitudes of the mean values of T_{1-10} increase with decreasing Le. In Figure 5, the positive contributions of the terms in the positively curved locations tend to increase the convexity of the flame surface, whereas the negative contributions of the terms act to reduce the convexity at $\kappa_m > 0$. By the same token, the positive contributions of the terms in the negatively curved locations tend to decrease the concavity of the flame surface, whereas the negative contributions of the terms act to increase the concavity at $\kappa_m < 0$.

It is evident from Figure 5 that the mean contribution of $T_1 = [(a_N n_{i,i})/2]$ assumes negative values for $\kappa_m < 0$ and positive values for $\kappa_m > 0$ for all cases on the $c = 0.8$ isosurface. In the reactive region a_N assumes predominantly positive values due to predominant preferential alignment between ∇c with the eigenvector associated with e_α, and this leads to negative (positive) values of $T_1 = [(a_N n_{i,i})/2]$ at the negatively (positively) curved zones on the flame surface.

In all cases, the term T_2 exhibits negative mean values associated with $\kappa_m > 0$ and positive mean values are obtained for $\kappa_m < 0$. The flow divergence ahead of the positively curved zones and flow convergence ahead of the negative curved regions lead to positive mean values of T_2 for $\kappa_m < 0$ and negative values of T_2 for $\kappa_m > 0$. The mean contribution of the flow stretching term $T_3 = -(S_{ij}n_{j,i})$ shows large positive values for convex topologies and weakly positive mean values for topologies with concave curvatures in the $Le = 1.0$ and 1.2 cases, whereas large positive mean values of T_3 are obtained at both highly positive and negative curved locations for the $Le = 0.8$ case. This is in agreement with Figure 4, which shows that the mean value of T_3 conditional on c remains positive throughout the flame. The probability of finding both cup concave and saddle concave topologies is greater in the $Le = 0.8$ case than in the $Le = 1.0$ and 1.2 cases, where T_3 can assume positive values, and thus the mean value of T_3 conditional on κ_m assumes large positive values in the negative curved zones. The mean value of T_4 conditional on κ_m shows an opposite behaviour in comparison to T_2, and thus it exhibits positive (negative) mean values for $\kappa_m > 0$ ($\kappa_m < 0$). The mean value of T_5 conditioned on curvature shows almost similar behavior to T_4 but the conditional mean value of T_5 remains negligible for $\kappa_m < 0$.

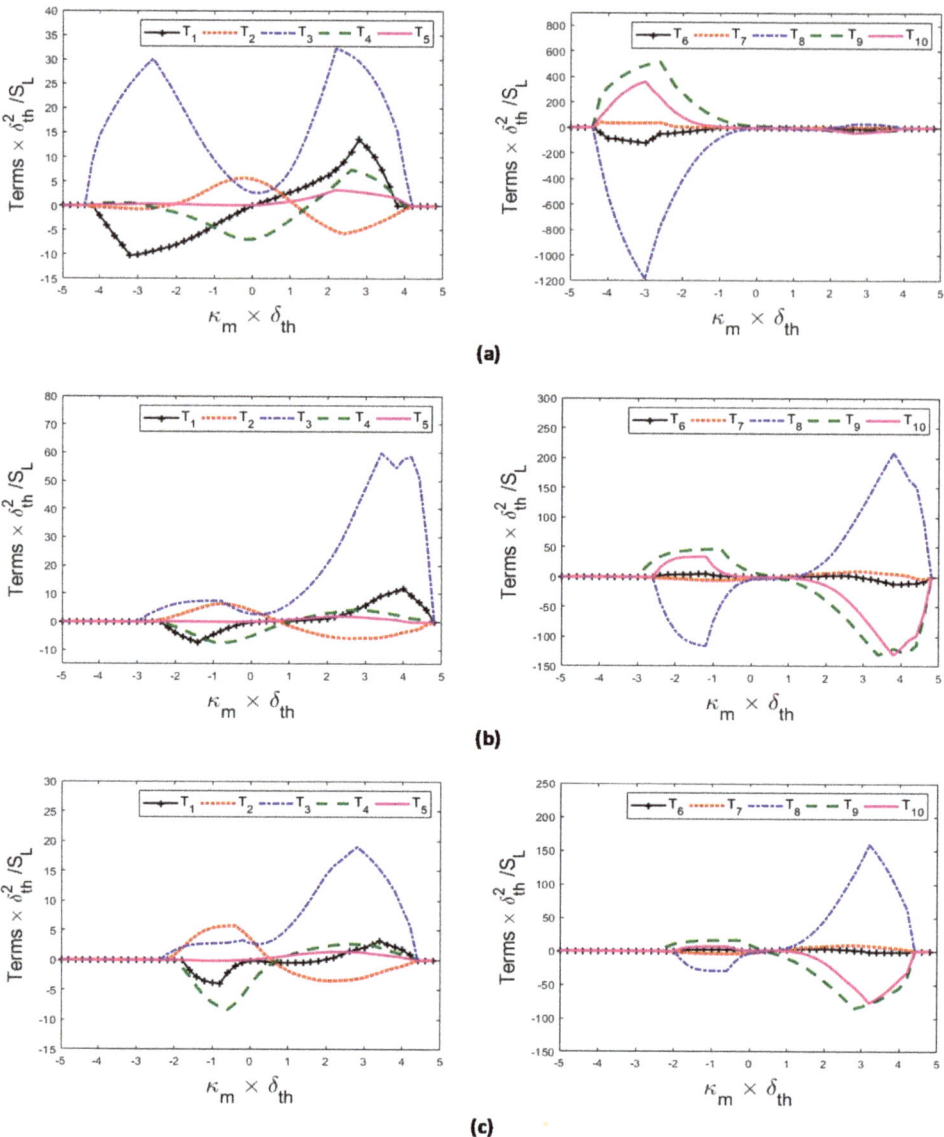

Figure 5. Profiles of the normalised mean values of the fluid flow induced terms (i.e., $T_{1-5} \times \delta_{th}^2/S_L$) (left column) and flame propagation induced terms (i.e., $T_{6-10} \times \delta_{th}^2/S_L$) (right column) of the curvature transport equation conditional upon normalised curvature $\kappa_m \times \delta_{th}$ on the $c = 0.8$ isosurface for cases (**a**–**c**) $Le = 0.8$, 1.0 and 1.2.

The mean contributions of the additional normal strain term $T_6 = \{(\partial S_d/\partial x_N)n_{i,i}/2\}$ and the term due to the normal gradient of the added normal strain rate T_7 conditional upon curvature remain negligible in comparison to the conditional mean values of T_8, T_9 and T_{10}. The mean contribution to the added stretching term, $T_8 = -(S_{ij}^a n_{j,i})$ conditioned on curvature plays a leading order role. It has already been mentioned that in the coordinate aligned with principal axes of curvature, T_8 can be expressed as: $T_8 = -S_d(\kappa_1^2 + \kappa_2^2)$, which implies a non-linear (e.g., cubic) curvature dependence of T_8 due to the curvature dependence of displacement speed S_d [32–41]. Consequently in all cases considered here, T_8 exhibits an asymmetric trend with respect to $\kappa_m = 0$, where its negative (positive) mean values correlating to $\kappa_m < 0$ ($\kappa_m > 0$), with $\kappa_m = 0$ being the inflection point for the $Le = 0.8$, 1.0 and 1.2 cases. However, in the $Le = 0.8$ case, the negative mean contribution of T_8 conditioned on κ_m for negatively curved regions remains much greater than the positive mean contribution in the positively curved zones. By contrast, the positive mean contribution of T_8 conditioned on κ_m for positively curved regions remains much greater than the negative mean contribution in the negatively curved zones in the $Le = 1.0$ and 1.2 cases. The mean contribution of the term T_9 due to the added strain rate gradients conditional upon curvature κ_m exhibits large positive values for topologies with highly negative curvatures but its mean value assumes small negative values in the positively curved zones for the $Le = 1.0$ and 1.2 cases. In the $Le = 0.8$ case, the negative mean contribution of T_9 remains negligible at highly positively curved locations. The mean values of the added vorticity curl contribution T_{10} conditional upon κ_m exhibit similar qualitative trend as that of T_9.

4.4. Overall Behaviour of the Terms in the Curvature Transport Equation

Figure 6 (left column) shows the normalised mean values of net fluid flow contributions $(T_1 + \cdots + T_5) \times (\delta_{th}^2/S_L)$, flame propagation induced added contributions $(T_6 + \cdots + T_{10}) \times (\delta_{th}^2/S_L)$ and the total contribution $(T_1 + \cdots + T_{10}) \times (\delta_{th}^2/S_L)$ conditional upon the reaction progress variable c. As T_1 and T_5 have negligible contributions and T_4 nullifies T_2, the net mean values of fluid motion terms $(T_1 + \cdots + T_5) \times (\delta_{th}^2/S_L)$ follow the statistical trend of the curvature flow stretching term $T_3 \times (\delta_{th}^2/S_L)$ in all cases considered here. It is evident from Figure 6 (left column) that the net contribution of the added terms due to flame propagation $(T_6 + \cdots + T_{10}) \times (\delta_{th}^2/S_L)$ dominates over the net contribution of the terms $(T_1 + \cdots + T_5) \times (\delta_{th}^2/S_L)$ arising from fluid flow. The mean contributions of $(T_6 + \cdots + T_{10}) \times (\delta_{th}^2/S_L)$ and $(T_1 + \cdots + T_{10}) \times (\delta_{th}^2/S_L)$ remain negative throughout the flame front for all cases. This behaviour originates because the negative contributions of T_8 and T_9 dominate over the positive contribution arising from T_7 on the unburned gas side, whereas the combined negative contributions of T_7 and T_8 dominate over the positive contribution of T_9 on the product side of the flame front.

The statistical behaviours of the normalised mean values of flow contributions $(T_1 + \cdots + T_5) \times (\delta_{th}^2/S_L)$, flame propagation induced contributions $(T_6 + \cdots + T_{10}) \times (\delta_{th}^2/S_L)$ and the total contribution $(T_1 + \cdots + T_{10}) \times (\delta_{th}^2/S_L)$ to the curvature transport, conditioned on curvature κ_m are shown in Figure 6 (right column) for the $c = 0.8$ isosurface. It is clear from Figure 6 (right column) that the mean values of the flow contributions $(T_1 + \cdots + T_5)$ are negligible in comparison to the mean values of the overall added flame propagation terms $(T_6 + \cdots + T_{10})$ in the negative curved locations, whereas the flow terms $(T_1 + \cdots + T_5)$ dominate over $(T_6 + \cdots + T_{10})$ for positive curvatures in the $Le = 0.8$ case. However, in the $Le = 1.0$ and 1.2 cases, the net mean flame propagation contribution $(T_6 + \cdots + T_{10})$ dominates over the mean contribution of the flow terms $(T_1 + \cdots + T_5)$ especially for the highly positive and negative curvatures. It can clearly be seen that the net contribution of the added terms induced by flame propagation conditional upon mean curvature κ_m shows the same trend of the added strain rate and vorticity gradients terms T_9 and T_{10} in the positively curved locations, whereas the added curvature stretching term T_8 determines the net mean behaviour of the added flame propagation contribution in the negatively curved zones. Thus, the net mean contribution of $(T_6 + \cdots + T_{10})$ remains negative for both highly negative and positive curvatures. However, the magnitudes of mean negative $(T_6 + \cdots + T_{10})$ in the positively curved zones are much greater than that in the negative curvatures in the $Le = 1.0$ and 1.2 cases. Just the opposite behaviour can be seen for the $Le = 0.8$ case, and the magnitude of the negative mean contribution of $(T_6 + \cdots + T_{10})$ at the negatively curved locations is especially large in this case due to large negative values of T_8. It is recalled that positivity (negativity) of $(T_1 + \cdots + T_{10})$ leads to increasing positivity (negativity) of mean curvature. As a result of such an amplification of positive (negative) curvature values, the position of flame elements in the diagrams shown in the right column of Figure 6 moves towards the right (left) and the opposite happens for a damping of curvature values. In other words, Figure 6 implies a characteristic motion of the location of flame elements in this diagram until they either behave in a neutral manner or small-scale wrinkles with characteristic length scales considerably smaller than the inner cut-off scale are smoothed out by molecular diffusion effects.

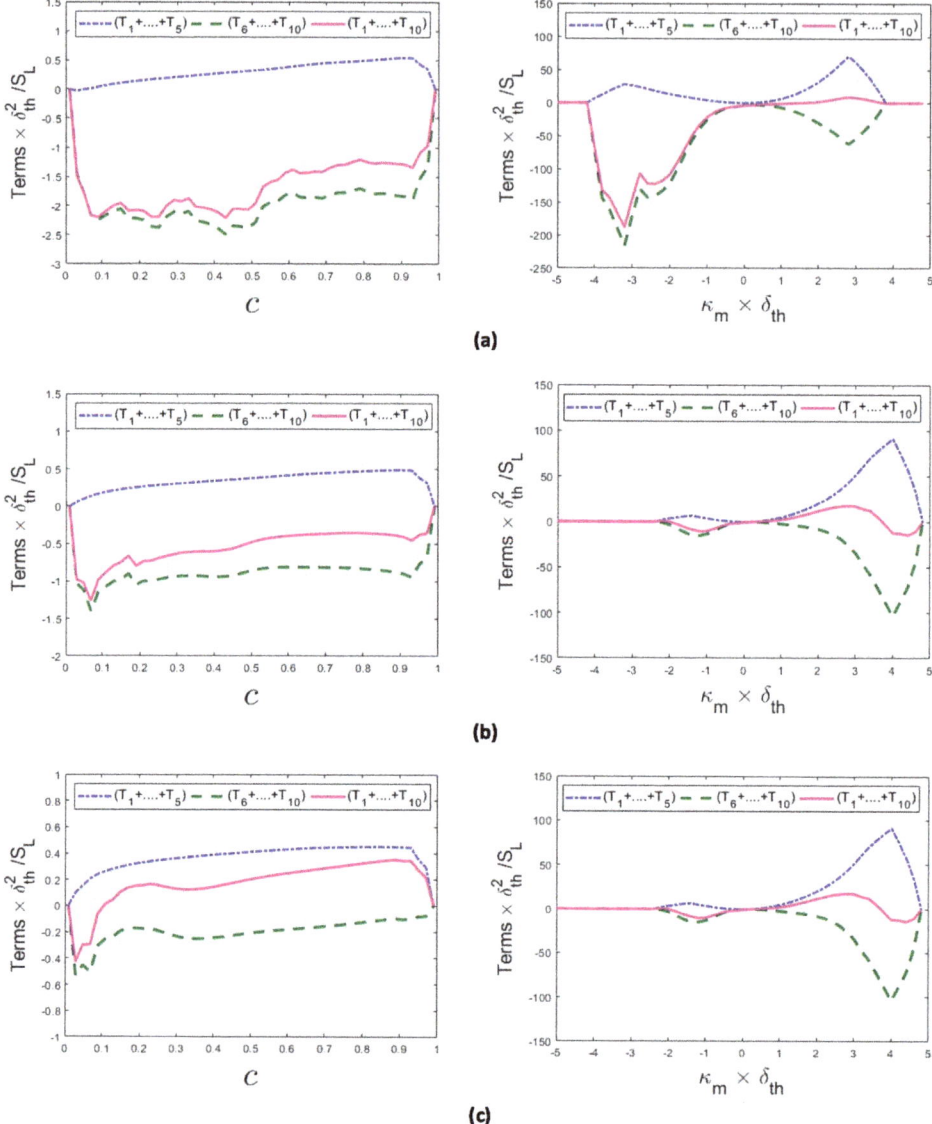

Figure 6. Profiles of normalised mean values of the flow contributions $(T_1 + \cdots + T_5) \times (\delta_{th}^2/S_L)$, flame propagation induced contributions $(T_6 + \cdots + T_{10}) \times (\delta_{th}^2/S_L)$ and the total contribution $(T_1 + \cdots + T_{10}) \times (\delta_{th}^2/S_L)$ to the curvature transport, conditioned on the reaction progress variable c (left column) and normalised curvature $\kappa_m \times \delta_{th}$ on the $c = 0.8$ isosurface (right column) for cases (**a**–**c**) $Le = 0.8$, 1.0 and 1.2.

Figure 6 (right column) shows that the mean contribution of $(T_1 + \cdots + T_{10})$ remains strongly negative at negative values of κ_m and mildly positive at positive values of κ_m in the $Le = 0.8$ case. However, in the $Le = 1.0$ and 1.2 cases, the mean contribution of $(T_1 + \cdots + T_{10})$ assumes negative values at highly positive and negative curved locations and positive values are obtained at moderately positive curved locations. The above findings indicate that the concavity of the negatively curved cusps, and the convexity of the positively curved bulges are promoted by the curvature transport in the $Le = 0.8$ case, which is further aided by the presence of high and low temperature (and thus burning rate) regions at positively and negatively curved regions, respectively. As a result, positively curved bulges with large radii of curvature and intermediate negatively curved cusps with small radii of curvature are likely to be observed in the $Le = 0.8$ case, which is consistent with the observations made from Figure 1 and the expected picture of the flame surface associated with $Le < 1$. In the $Le = 1.0$ and 1.2 cases, the negative mean contribution of $(T_1 + \cdots + T_{10})$ acts to reduce the convexity of the positively curved bulges. By contrast, the negative mean $(T_1 + \cdots + T_{10})$ values at negatively curved locations in the $Le = 1.0$ and 1.2 cases act to increase the concavity of the negatively curved cusps. However, these structures are unstable and are eventually smoothed out due to increased S_d at negatively curved locations. This effect is particularly strong in the $Le = 1.2$ case because of large temperature (and thus also burning rate) at the negatively curved regions.

4.5. Relations of the Terms of the Curvature Transport Equation with Local Curvature

The mean values of $T_{1-5} \times (\delta_{th}^2/S_L)$ conditional on $\kappa_m \times \delta_{th}$ and $\kappa_g \times \delta_{th}^2$ for the $Le = 0.8$, 1.0 and 1.2 cases are shown in Figure 7 for the $c = 0.8$ isosurface. The corresponding variations of $T_{6-10} \times (\delta_{th}^2/S_L)$ conditional on $\kappa_m \times \delta_{th}$ and $\kappa_g \times \delta_{th}^2$ on the $c = 0.8$ isosurface are shown in Figure 8 for these cases. Figure 7 indicates high magnitudes of T_1 and T_3 are mostly obtained for concave topologies on the flame surface for the $Le = 0.8$ case. The same trend is seen for T_1, T_2 and T_4 in the $Le = 1.0$ and 1.2 flames. The large magnitudes of T_5 are associated with the cup convex flame topologies in all cases, and the magnitudes of these terms increase with decreasing Lewis number.

It is evident from Figure 8 that the terms T_{6-10} in the $Le = 0.8$ case show high magnitudes for cup concave (i.e., $\kappa_m < 0$ and $\kappa_g > 0$) and saddle concave (i.e., $\kappa_m < 0$ and $\kappa_g < 0$) topologies. However, large magnitudes of these terms are mostly obtained for cup convex and saddle convex flame topologies in the $Le = 1.0$ and 1.2 cases. Moreover, large negative magnitudes of T_8 and large positive magnitudes of T_9 and T_{10} can be discerned for small (both positive and negative) values of κ_g for weakly negatively curved zones (i.e., $\kappa_m < 0$) in the $Le = 1.0$ case.

4.6. Modelling Implications

It can be seen from Figures 4–8 that the net flame propagation contribution to the curvature κ_m transport plays significant roles for all flames considered here. This suggests that displacement speed S_d is not only interlinked with curvature (i.e., a negative correlation exists between S_d and κ_m [32–41]) but it also affects flame wrikling through the curvature evolution. Moreover, it was demonstrated elsewhere [67,69,83–86] that the curvature κ_m and its interrelation with displacement speed S_d plays a key role in the Flame Surface Density (FSD) and scalar dissipation rate (SDR) transports. Furthermore, the analysis of curvature evolution also reveals that the displacement speed induced terms are principally responsible for the generation of negatively curved cusps in the flames with $Le < 1$.

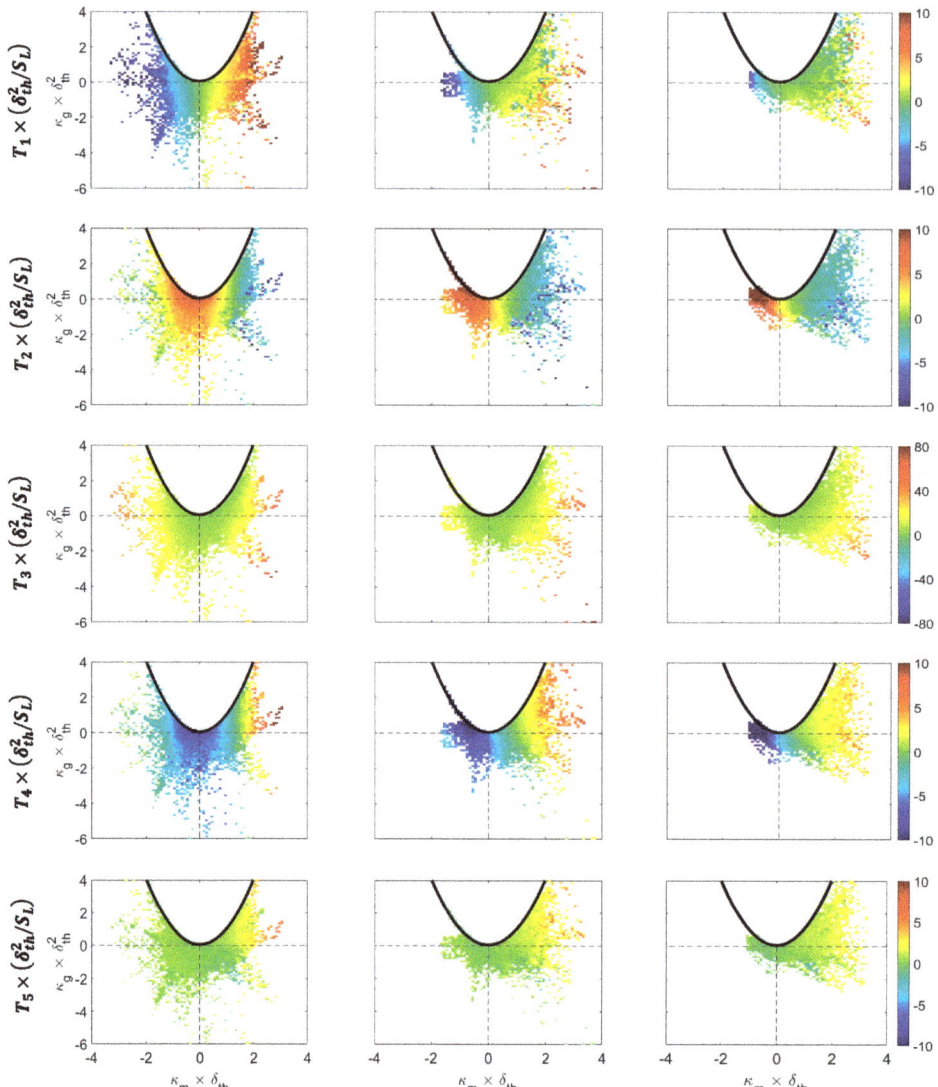

Figure 7. Variations of the mean values of $T_{1-5} \times (\delta_{th}^2/S_L)$ (1st–5th row) conditional on $\kappa_m \times \delta_{th}$ and $\kappa_g \times \delta_{th}^2$ for Le = 0.8, 1.0 and 1.2 flames (1st–3rd column) for $c = 0.8$ isosurface.

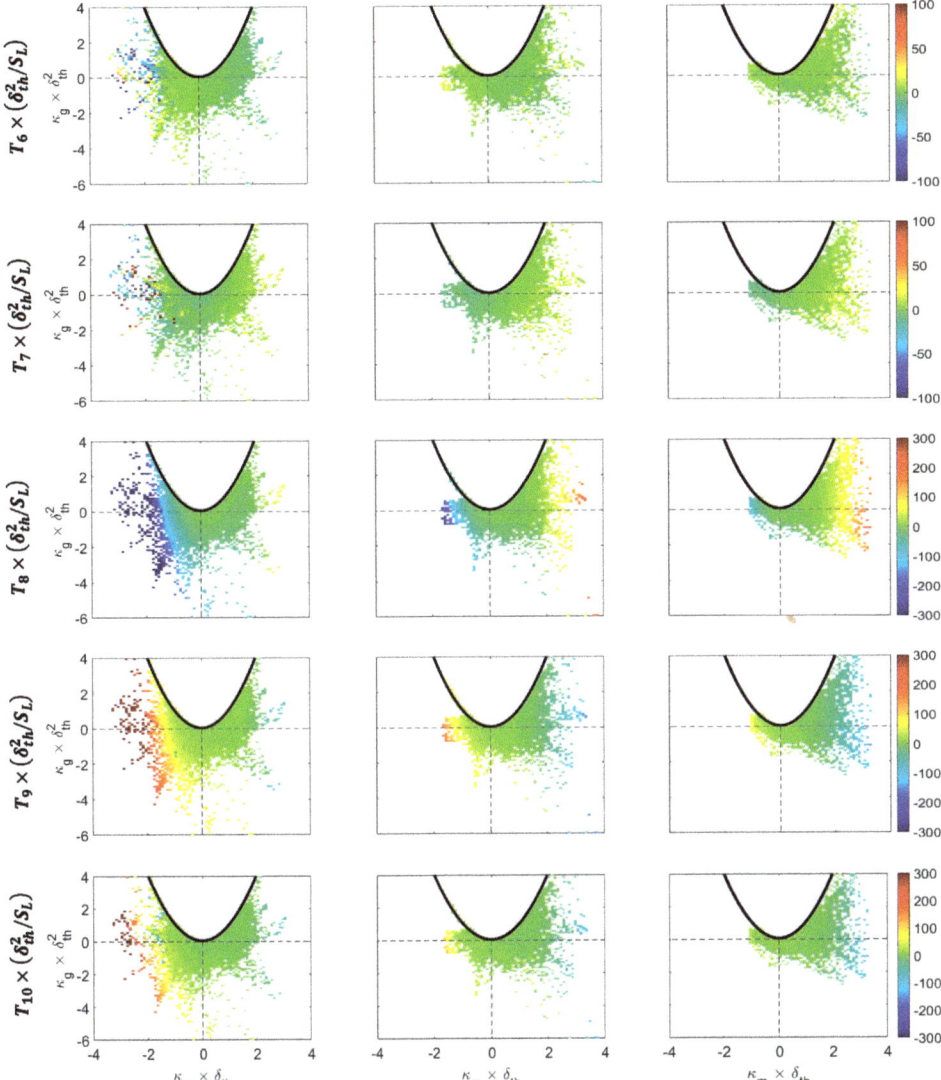

Figure 8. Variations of the mean values of $T_{6-10} \times (\delta_{th}^2/S_L)$ (1st–5th row) conditional on $\kappa_m \times \delta_{th}$ and $\kappa_g \times \delta_{th}^2$ for $Le = 0.8$, 1.0 and 1.2 flames (1st–3rd column) for $c = 0.8$ isosurface.

5. Conclusions

The effects of characteristic Lewis number on the statistical behaviours of the different terms of the curvature transport equation have been analysed based on three-dimensional compressible DNS data of spherically expanding turbulent premixed flames with $Le = 0.8$, 1.0 and 1.2. The statistically spherical flames with $Le = 0.8$, 1.0 and 1.2 had the same initial radius before they were allowed to interact with initially homogeneous isotropic decaying turbulence. It has been found that the flame surface area and volume-integrated burning rate increase with decreasing Le, which is consistent with several previous findings. The greater extent of flame wrinkling in the $Le = 0.8$ case is reflected in the wider range of both positive and negative curvatures than in the corresponding $Le = 1.0$ and 1.2

cases where the joint PDF remains almost symmetrical about $\kappa_m = 0$ on the reactant side, but skews gradually toward positive values of κ_m in the reaction and hot product zones. The PDFs of curvature for the $Le = 0.8$ case show higher probabilities of finding sharply negatively curved cusps than in the corresponding $Le = 1.0$ and 1.2 cases. Moreover, the saddle topologies have been found to be more frequent in the $Le = 0.8$ case than in the other cases considered in this analysis.

It has been found that the mean contributions of flame normal gradient of normal strain rate and the added strain rate due to flame displacement speed (i.e., T_2 and T_9) assume negative values (i.e., promote concavity of the flame surface) towards the unburned gas and positive values (i.e., promote convexity of the flame surface) on the burned gas side of the flame. By contrast, the mean contributions arising from flow strain rate gradient and the flame normal gradient of the added strain rate (i.e., T_4 and T_7) tend to promote positive and negative curvatures on reactant and product sides of the flame. The mean added curvature stretch term T_8 exhibits negative mean values throughout the flame, while the curvature flow stretching term T_3 remains positive throughout the flame front.

The mean added curvature stretch term T_8 conditional on curvature shows different behaviour in response to the changes in Le. The added curvature stretch term T_8 shows negative mean values for $\kappa_m < 0$ and negligible mean value of T_8 is obtained for $\kappa_m > 0$ in the $Le = 0.8$ case but positive (negative) mean values of T_8 are obtained for positive (negative) κ_m values in the $Le = 1.0$ and 1.2 cases. The mean value of curvature flow stretching term T_3 remains positive for both $\kappa_m > 0$ and $\kappa_m < 0$ but the magnitude of T_3 associated with $\kappa_m < 0$ increases significantly with decreasing Le. The terms due to normal gradient of added strain rate T_9 and the curl of added vorticity T_{10} assume positive (negative) values for negative (positive) κ_m values. The net mean contribution of the terms arising from flame propagation (i.e., $(T_6 + \cdots + T_{10})$) dominates over the net mean contribution of the terms due to background fluid motion (i.e., $(T_1 + \cdots + T_5)$) for the negatively curved locations but the opposite behaviour has been observed for the positively curved zones in the $Le = 0.8$ case. However, in the $Le = 1.0$ and 1.2 cases the net mean flame propagation contribution $(T_6 + \cdots + T_{10})$ dominates over the mean contribution of the flow terms $(T_1 + \cdots + T_5)$ especially for high magnitudes of κ_m. For the $Le = 1.0$ and 1.2 cases, the net mean contribution of $(T_1 + \cdots + T_{10})$ remains negative for high positive curvatures but this contribution assumes weak negative values for negatively curved regions. By contrast, in the $Le = 0.8$ case, the net mean contribution of $(T_1 + \cdots + T_{10})$ assumes negative values for $\kappa_m < 0$ and mild positive values are obtained for $\kappa_m > 0$, which is indicative of promoting sharply negatively curved cusps and positively curved bulges with large radii of curvature. This tendency is further augmented by high temperatures (also burning rates) in the positively curved and low temperatures (also burning rates) in the negatively curved regions in the $Le = 0.8$ case, which is expected in the presence of thermo-diffusive instability. By contrast, highly positively curved bulges are not promoted in the $Le = 1.0$ and 1.2 cases, and weak negative mean contributions of $(T_1 + \cdots + T_{10})$ at the negatively curved zones tend to produce negatively curved cusps which are eventually smoothed by large values of S_d in these regions. Thus, flame propagation tends to smoothen the flame wrinkles induced by turbulence in the $Le = 1.0$ and 1.2 cases and this effect is stronger in the $Le = 1.2$ case due to high temperature (and thus high reaction rate) at the negatively curved regions. Furthermore, it has been found that flame propagation plays a pivotal role in the curvature evolution irrespective of the characteristic Lewis number and thus the interrelation between displacement speed and curvature needs to be explicitly accounted for in the context of FSD and SDR closures, especially in order to predict thermo-diffusive instability effects for $Le < 1$ flames. This model development along with further analysis of curvature evolution using detailed chemistry and transport-based DNS data will form the basis of further investigations.

Author Contributions: Conceptualization, N.C. and M.K.; simulation, M.K.; Postprocessing code development, A.A., N.C.; formal analysis, A.A., N.C.; writing—original draft preparation, A.A., N.C.; writing—review and editing, N.C., M.K.; visualization, A.A., N.C. M.K.; supervision, N.C.; funding acquisition, N.C., M.K.

Funding: This research was funded by Engineering and Physical Sciences Research Council (EPSRC) (EP/K025163/1) and the German Research Foundation (DFG, KL1456/5-1).

Acknowledgments: We received computational support from ARCHER, Rocket High Performance Computing and Gauss Centre for Supercomputing (grant: pn69ga).

Conflicts of Interest: The authors declare no conflict of interest.

References

1. Abdel-Gayed, R.G.; Al-Khishali, K.J.; Bradley, D. Turbulent burning velocities and flame straining in explosions. *Proc. R. Soc. Lond. A* **1984**, *391*, 393–414. [CrossRef]
2. Beretta, G.P.; Rashidi, M.; Keck, J.C. Turbulent flame propagation and combustion in spark ignition engines. *Combust. Flame* **1983**, *52*, 217–245. [CrossRef]
3. Bradley, D.; Gaskell, P.H.; Gu, X.J. Burning velocities, Markstein lengths, and flame quenching for spherical methane-air flames: A computational study. *Combust. Flame* **1996**, *104*, 176–198. [CrossRef]
4. Renou, B.; Boukhalfa, A.; Puechberty, D.; Trinit'e, M. Local scalar flame properties of freely propagating premixed turbulent flames at various Lewis numbers. *Combust. Flame* **2000**, *123*, 507–521. [CrossRef]
5. Haq, M.Z.; Sheppard, C.G.W.; Woolley, R.; Greenhalgh, D.A.; Lockett, R.D. Wrinkling and curvature of laminar and turbulent premixed flames. *Combust. Flame* **2002**, *131*, 1–15. [CrossRef]
6. Bradley, D.; Haq, M.Z.; Hicks, R.A.; Kitagawa, T.; Lawes, M.; Sheppard, C.G.W.; Woolley, R. Turbulent burning velocity, burned gas distribution, and associated flame surface definition. *Combust. Flame* **2002**, *133*, 415–430. [CrossRef]
7. Lawes, M.; Ormsby, M.P.; Sheppard, C.G.W.; Woolley, R. The turbulent burning velocity of iso-octane/air mixtures. *Combust. Flame* **2012**, *159*, 1949–1959. [CrossRef]
8. Chaudhuri, S.; Wu, F.; Zhu, D.; Law, C.K. Flame speed and self-similar propagation of expanding turbulent premixed flames. *Phys. Rev. Lett.* **2012**, *108*, 044503. [CrossRef]
9. Akkerman, V.; Chaudhuri, S.; Law, C.K. Accelerative propagation and explosion triggering by expanding turbulent premixed flames. *Phys. Rev. E* **2013**, *87*, 23008. [CrossRef]
10. Pope, S.B.; Cheng, W.K. Statistical calculations of spherical turbulent flames. *Proc. Combust. Inst.* **1986**, *21*, 1473–1481.
11. Baum, M.; Poinsot, T. Effects of mean flow on premixed flame ignition. *Combust. Sci. Technol.* **1995**, *106*, 19–39. [CrossRef]
12. Poinsot, T.; Candel, S.; Trouve, A. Applications of direct numerical simulation to premixed turbulent combustion. *Prog. Energy Combust. Sci.* **1995**, *21*, 531–576. [CrossRef]
13. Schmid, H.-P.; Habisreuther, P.; Leuckel, W. A model for calculating heat release in premixed turbulent flames. *Combust. Flame* **1998**, *113*, 79–91. [CrossRef]
14. Nwagwe, I.K.; Weller, H.G.; Tabor, G.R.; Gosman, A.D.; Lawes, M.; Sheppard, C.G.W.; Wooley, R. Measurements and large eddy simulations of turbulent premixed flame kernel growth. *Proc. Combust. Inst.* **2000**, *28*, 59–65. [CrossRef]
15. Jenkins, K.W.; Cant, R.S. Curvature effects on flame kernels in a turbulent environment. *Proc. Combust. Inst.* **2002**, *29*, 2023–2029. [CrossRef]
16. Tabor, G.; Weller, H.G. Large eddy simulation of premixed turbulent combustion using flame surface wrinkling model. *Flow Turbul. Combust.* **2004**, *72*, 1–27. [CrossRef]
17. van Oijen, J.A.; Groot, G.R.A.; Bastiaans, R.J.M.; de Goey, L.P.H. A flamelet analysis of the burning velocity of premixed turbulent expanding flames. *Proc. Combust. Inst.* **2005**, *30*, 657–664. [CrossRef]
18. Gashi, S.; Hult, J.; Jenkins, K.W.; Chakraborty, N.; Cant, R.S.; Kaminski, C.F. Curvature and Wrinkling of Premixed Flame Kernels-Comparisons of OH PLIF and DNS data. *Proc. Combust. Inst.* **2005**, *30*, 809–817. [CrossRef]
19. Thevenin, D. Three-dimensional direct simulations and structure of expanding turbulent methane flames. *Proc. Combust. Inst.* **2005**, *30*, 629–637. [CrossRef]
20. Klein, M.; Chakraborty, N.; Jenkins, K.W.; Cant, R.S. Effects of initial radius on the propagation of premixed flame kernels in a turbulent environment. *Phys. Fluids* **2006**, *18*, 055102. [CrossRef]
21. Jenkins, K.W.; Klein, M.; Chakraborty, N.; Cant, R.S. Effects of strain rate and curvature on the propagation of a spherical flame kernel in the thin-reactionzones regime. *Combust. Flame* **2006**, *145*, 415–434. [CrossRef]
22. Klein, M.; Chakraborty, N.; Cant, R.S. Effects of turbulence on self-sustained combustion in premixed flame kernels: A direct numerical simulation (DNS) study. *Flow Turbul. Combust.* **2008**, *81*, 583–607. [CrossRef]

23. Hult, J.; Gashi, S.; Chakraborty, N.; Klein, M.; Jenkins, K.W.; Cant, R.S.; Kaminski, C. Measurement of Flame Surface Density for Turbulent Premixed Flames using PLIF and DNS. *Proc. Combust. Inst.* **2007**, *31*, 1319–1326. [CrossRef]
24. Chakraborty, N.; Klein, M.; Cant, R.S. Stretch rate effects on displacement speed in turbulent premixed flame kernels in the thin reaction zones regime. *Proc. Combust. Inst.* **2007**, *31*, 1385–1392. [CrossRef]
25. Dunstan, T.D.; Jenkins, K.W. Flame surface density distribution in turbulent flame kernels during the early stages of growth. *Proc. Combust. Inst.* **2009**, *32*, 1427–1434. [CrossRef]
26. Dunstan, T.D.; Jenkins, K.W. The effects of hydrogen substitution on turbulent premixed methane-air kernels using direct numerical simulation. *Int. J. Hydrogen Energy* **2009**, *34*, 8389–8404. [CrossRef]
27. Chakraborty, N.; Rogerson, J.W.; Swaminathan, N. The scalar gradient alignment statistics of flame kernels and its modelling implications for turbulent premixed combustion. *Flow Turbul. Combust.* **2010**, *85*, 25–55. [CrossRef]
28. Lecocq, G.; Richard, S.; Colin, O.; Vervisch, L. Hybrid presumed PDF and flame surface density approaches for large-eddy simulation of premixed turbulent combustion. part 2: Early flame development after sparking. *Combust. Flame* **2011**, *158*, 1215–1226. [CrossRef]
29. Colin, O.; Truffin, K. A spark ignition model for large eddy simulation based on an FSD transport equation (ISSIM-LES). *Proc. Combust. Inst.* **2011**, *33*, 3097–3104. [CrossRef]
30. Ahmed, I.; Swaminathan, N. Simulation of spherically expanding turbulent premixed flames. *Combust. Sci.Technol.* **2013**, *185*, 1509–1540. [CrossRef]
31. Ahmed, I.; Swaminathan, N. Simulation of turbulent explosion of hydrogen-air mixtures. *Int. J. Hydrogen Energy* **2014**, *39*, 9562–9572. [CrossRef]
32. Peters, N.; Terhoeven, P.; Chen, J.H.; Echekki, T. Statistics of flame displacement speeds from computations of 2-D unsteady methane-air flames. *Proc. Combust. Inst.* **1998**, *27*, 833–839. [CrossRef]
33. Echekki, T.; Chen, J.H. Analysis of the contributions of curvature to premixed flame propagation. *Combust. Flame* **1999**, *118*, 308–311. [CrossRef]
34. Chen, J.H.; Im, H.G. Correlation of flame speed with stretch in turbulent premixed Methane/Air flames. *Proc. Combust. Inst.* **1998**, *27*, 819–826. [CrossRef]
35. Chen, J.H.; Im, H.G. Stretch effects on the burning velocity of turbulent premixed Hydrogen/Air flames. *Proc. Combust. Inst.* **2000**, *28*, 211–218. [CrossRef]
36. Hawkes, E.R.; Chen, J.H. Direct numerical simulation of hydrogen-enriched lean premixed methane air flames. *Combust. Flame* **2004**, *138*, 242–258. [CrossRef]
37. Hawkes, E.R.; Chen, J.H. Evaluation of models for flame stretch due to curvature in the thin reaction zones regime. *Proc. Combust. Inst.* **2005**, *30*, 647–655. [CrossRef]
38. Chakraborty, N.; Cant, S. Unsteady effects of strain rate and curvature on turbulent premixed flames in inlet-outlet configuration. *Combust. Flame* **2004**, *137*, 129–147. [CrossRef]
39. Chakraborty, N.; Cant, R.S. Influence of Lewis number on curvature effects in turbulent premixed flame propagation in the thin reaction zones regime. *Phys. Fluids* **2005**, *17*, 105105. [CrossRef]
40. Chakraborty, N. Comparison of displacement speed statistics of turbulent premixed flames in the regimes representing combustion in corrugated flamelets and thin reaction zones. *Phys. Fluids* **2007**, *19*, 105109. [CrossRef]
41. Han, I.; Huh, K.Y. Roles of displacement speed on evolution of flame surface density for different turbulent intensities and Lewis numbers in turbulent premixed combustion. *Combust. Flame* **2008**, *152*, 194–205. [CrossRef]
42. Rutland, C.; Trouvé, A. Direct Simulations of premixed turbulent flames with nonunity Lewis numbers. *Combust. Flame* **1993**, *94*, 41–57. [CrossRef]
43. Chakraborty, N.; Cant, R.S. Effects of strain rate and curvature on surface density function transport in turbulent premixed flames in the thin reaction zones regime. *Phys. Fluids* **2005**, *17*, 065108. [CrossRef]
44. Chakraborty, N.; Klein, M. Influence of Lewis number on the Surface Density Function transport in the thin reaction zones regime for turbulent premixed flames. *Phys. Fluids* **2008**, *20*, 065102. [CrossRef]
45. Chakraborty, N.; Klein, M. Effects of global flame curvature on surface density function transport in turbulent premixed flame kernels in the thin reaction zone regime. *Proc. Combust. Inst.* **2009**, *32*, 1435–1443. [CrossRef]

46. Sandeep, A.; Proch, F.; Kempf, A.M.; Chakraborty, N. Statistics of strain rates and Surface Density Function in a flame-resolved high-fidelity simulation of a turbulent premixed bluff body burner. *Phys. Fluids* **2018**, *30*, 065101. [CrossRef]
47. Chakraborty, N.; Hawkes, E.R.; Chen, J.H.; Cant, R.S. Effects of strain rate and curvature on Surface Density Function transport in turbulent premixed CH_4-air and H_2-air flames: A comparative study. *Combust. Flame* **2008**, *154*, 259–280. [CrossRef]
48. Klein, M.; Alwazzan, D.; Chakraborty, N. A Direct Numerical Simulation analysis of pressure variation in turbulent premixed Bunsen burner flames-Part 1: Scalar gradient and strain rate statistics. *Comput. Fluids* **2018**. [CrossRef]
49. Klein, M.; Alwazzan, D.; Chakraborty, N. A Direct Numerical Simulation analysis of pressure variation in turbulent premixed Bunsen burner flames-Part 2: Surface Density Function transport statistics. *Comput. Fluids* **2018**. [CrossRef]
50. Reddy, H.; Abraham, J. Two-dimensional direct numerical simulation evaluation of the flame-surface density model for flames developing from an ignition kernel in lean methane/air mixtures under engine conditions. *Phys. Fluids* **2012**, *24*, 105108. [CrossRef]
51. Pope, S.B. The evolution of surfaces in turbulence. *Int. J. Eng. Sci.* **1988**, *26*, 445–469. [CrossRef]
52. Dopazo, C.; Martin, J.; Cifuentes, L.; Hierro, J. Strain, rotation and curvature of non-material propagating iso-scalar surfaces in homogeneous turbulence. *Flow Turbul. Combust.* **2018**. [CrossRef]
53. Cifuentes, L.; Dopazo, C.; Karichedu, A.; Chakraborty, N.; Kempf, A.M. Analysis of flame curvature evolution in a turbulent premixed bluff Body burner. *Phys. Fluids* **2018**, *30*, 095101. [CrossRef]
54. Sivashinsky, G.I. Diffusional-thermal theory of cellular flames. *Combust. Sci. Technol.* **1977**, *16*, 137–146. [CrossRef]
55. Clavin, P.; Williams, F.A. Effects of molecular diffusion and thermal expansion on the structure and dynamics of turbulent premixed flames in turbulent flows of large scale and small intensity. *J. Fluid Mech.* **1982**, *128*, 251–282. [CrossRef]
56. Ashurst, W.T.; Peters, N.; Smooke, M.D. Numerical simulation of turbulent flame structure with non-unity Lewis number. *Combust. Sci. Technol.* **1987**, *53*, 339–375. [CrossRef]
57. Haworth, D.C.; Poinsot, T.J. Numerical simulations of Lewis number effects in turbulent premixed flames. *J. Fluid Mech.* **1992**, *244*, 405–436. [CrossRef]
58. Trouvé, A.; Poinsot, T. The evolution equation for flame surface density in turbulent premixed combustion. *J. Fluid Mech.* **1994**, *278*, 1–31. [CrossRef]
59. Dopazo, C.; Cifuentes, L.; Chakraborty, N. Vorticity budgets in premixed combusting turbulent flows at different Lewis numbers. *Phys. Fluids* **2017**, *29*, 045106. [CrossRef]
60. Dopazo, C.; Cifuentes, L.; Martin, J.; Jimenez, C. Strain rates normal to approaching isoscalar surfaces in a turbulent premixed flame. *Combust. Flame* **2015**, *162*, 1729–1736. [CrossRef]
61. Dopazo, C.; Cifuentes, L.; Alwazzan, D.; Chakraborty, N. Influence of the Lewis number on effective strain rates in weakly turbulent premixed combustion. *Combust. Sci. Technol.* **2018**, *190*, 591–614. [CrossRef]
62. Chakraborty, N.; Kolla, H.; Sankaran, R.; Hawkes, E.R.; Chen, J.H.; Swaminathan, N. Determination of three-dimensional quantities related to scalar dissipation rate and its transport from two-dimensional measurements: Direct Numerical Simulation based validation. *Proc. Combust. Inst.* **2013**, *34*, 1151–1162. [CrossRef]
63. Lipatnikov, A.N.; Nishiki, S.; Hasegawa, T. A direct numerical study of vorticity transformation in weakly turbulent premixed flames. *Phys. Fluids* **2014**, *26*, 105104. [CrossRef]
64. Gao, Y.; Chakraborty, N.; Klein, M. Assessment of sub-grid scalar flux modelling in premixed flames for Large Eddy Simulations: A-priori Direct Numerical Simulation. *Eur. J. Mech. Fluids-B* **2015**, *52*, 97–108. [CrossRef]
65. Papapostolou, V.; Wacks, D.H.; Klein, M.; Chakraborty, N.; Im, H.G. Enstrophy transport conditional on local flow topologies in different regimes of premixed turbulent combustion. *Sci. Rep.* **2017**, *7*, 11545. [CrossRef] [PubMed]
66. Klein, M.; Kasten, C.; Chakraborty, N.; Mukhadiyev, N.; Im, H.G. Turbulent scalar fluxes in Hydrogen-Air premixed flames at low and high Karlovitz numbers. *Combust. Theory Model.* **2018**, *22*, 1033–1048. [CrossRef]

67. Gao, Y.; Chakraborty, N. Modelling of Lewis Number dependence of Scalar dissipation rate transport for Large Eddy Simulations of turbulent premixed combustion. *Numer. Heat Trans. A* **2016**, *69*, 1201–1222. [CrossRef]
68. Lai, J.; Chakraborty, N. Effects of Lewis Number on head on quenching of turbulent premixed flame: A Direct Numerical Simulation analysis. *Flow Turbul. Combust.* **2016**, *96*, 279–308. [CrossRef]
69. Gao, Y.; Minamoto, Y.; Tanahashi, M.; Chakraborty, N. A priori assessment of scalar dissipation rate closure for Large Eddy Simulations of turbulent premixed combustion using a detailed chemistry Direct Numerical Simulation database. *Combust. Sci. Technol.* **2016**, *188*, 1398–1423. [CrossRef]
70. Lai, J.; Klein, M.; Chakraborty, N. Direct Numerical Simulation of head-on quenching of statistically planar turbulent premixed methane-air flames using a detailed chemical mechanism. *Flow Turbul. Combust.* **2018**, *101*, 1073–1091. [CrossRef]
71. Jenkins, K.W.; Cant, R.S. DNS of turbulent flame kernels. In *Proceedings of the 2nd AFOSR Conference on DNS and LES*; Liu, C., Sakell, L., Beautner, T., Eds.; Kluwer Academic Publishers: Dordrecht, The Netherlands, 1999; pp. 192–202.
72. Muppala, S.R.; Aluri, N.K.; Dinkelacker, F.; Leipertz, A. Development of an Algebraic Reaction rate approach for the Numerical Calculation of Turbulent Premixed Methane, Ethylene and Propane/air flames at Pressures up to 1.0 MPa. *Combust. Flame* **2005**, *140*, 257–266. [CrossRef]
73. Kobayashi, H.; Tamura, H.; Maruta, K.; Nikola, T.; Williams, F.A. Burning velocity of turbulent premixed flames in a high-pressure environment. *Proc. Combust. Inst.* **1996**, *26*, 389–396. [CrossRef]
74. Wray, A.A. *Minimal Storage Time Advancement Schemes for Spectral Methods*; NASA Ames Research Center: Mountain View, CA, USA, 1990; Unpublished work.
75. Poinsot, T.; Lele, S.K. Boundary conditions for direct simulation of compressible viscous flows. *J. Comp. Phys.* **1992**, *101*, 104–129. [CrossRef]
76. Rogallo, R.S. *Numerical Experiments in Homogeneous Turbulence*; NASA Technical Memorandum 81315; NASA Ames Research Center: Mountain View, CA, USA, 1981.
77. Pope, S.B. *Turbulent Flows*; Cambridge University Press: Cambridge, UK, 2000.
78. Peters, N. *Turbulent Combustion, Cambridge Monograph on Mechanics*; Cambridge University Press: Cambridge, UK, 2000.
79. Dopazo, C.; Martin, J.; Hierro, J. Local geometry of isoscalar surfaces. *Phys. Rev. E* **2007**, *76*, 056316. [CrossRef]
80. Hartung, G.; Hult, J.; Kaminski, C.F.; Rogerson, J.W.; Swaminathan, N. Effect of heat release on turbulence and scalar-turbulence interaction in premixed combustion. *Phys. Fluids* **2008**, *20*, 035110. [CrossRef]
81. Chakraborty, N.; Swaminathan, N. Influence of Damköhler number on turbulence-scalar interaction in premixed flames, Part I: Physical Insight. *Phys. Fluids* **2007**, *19*, 045103. [CrossRef]
82. Chakraborty, N.; Klein, M.; Swaminathan, N. Effects of Lewis number on reactive scalar gradient alignment with local strain rate in turbulent premixed flames. *Proc. Combust. Inst.* **2009**, *32*, 1409–1417. [CrossRef]
83. Chakraborty, N.; Cant, R.S. A-Priori Analysis of the curvature and propagation terms of the Flame Surface Density transport equation for Large Eddy Simulation. *Phys. Fluids* **2007**, *19*, 105101. [CrossRef]
84. Chakraborty, N.; Cant, R.S. Direct Numerical Simulation analysis of the Flame Surface Density transport equation in the context of Large Eddy Simulation. *Proc. Combust. Inst.* **2009**, *32*, 1445–1453. [CrossRef]
85. Gao, Y.; Chakraborty, N.; Swaminathan, N. Local strain rate and curvature dependences of scalar dissipation rate transport in turbulent premixed flames: A Direct Numerical Simulation analysis. *J. Combust.* **2014**, *2014*, 280671. [CrossRef]
86. Gao, Y.; Chakraborty, N.; Swaminathan, N. Scalar dissipation rate transport and its modelling for Large Eddy Simulations of turbulent premixed combustion. *Combust. Sci. Technol.* **2015**, *187*, 362–383. [CrossRef]

© 2019 by the authors. Licensee MDPI, Basel, Switzerland. This article is an open access article distributed under the terms and conditions of the Creative Commons Attribution (CC BY) license (http://creativecommons.org/licenses/by/4.0/).

Article

Investigation of the Turbulent Near Wall Flame Behavior for a Sidewall Quenching Burner by Means of a Large Eddy Simulation and Tabulated Chemistry

Arne Heinrich [1,*], Guido Kuenne [1], Sebastian Ganter [1], Christian Hasse [2] and Johannes Janicka [1]

[1] Institute of Energy and Power Plant Technology, TU Darmstadt, Otto-Berndt-Strasse 3, 64287 Darmstadt, Germany; kuenne@ekt.tu-darmstadt.de (G.K.); ganter@ekt.tu-darmstadt.de (S.G.); janicka@ekt.tu-darmstadt.de (J.J.)
[2] Institute Simulation of Reactive Thermo-Fluid Systems, TU Darmstadt, Otto-Berndt-Strasse 2, 64287 Darmstadt, Germany; hasse@stfs.tu-darmstadt.de
* Correspondence: heinrich@ekt.tu-darmstadt.de; Tel.: +49-6151-1628812

Received: 25 July 2018; Accepted: 3 September 2018; Published: 6 September 2018

Abstract: Combustion will play a major part in fulfilling the world's energy demand in the next 20 years. Therefore, it is necessary to understand the fundamentals of the flame–wall interaction (FWI), which takes place in internal combustion engines or gas turbines. The FWI can increase heat losses, increase pollutant formations and lowers efficiencies. In this work, a Large Eddy Simulation combined with a tabulated chemistry approach is used to investigate the transient near wall behavior of a turbulent premixed stoichiometric methane flame. This sidewall quenching configuration is based on an experimental burner with non-homogeneous turbulence and an actively cooled wall. The burner was used in a previous study for validation purposes. The transient behavior of the movement of the flame tip is analyzed by categorizing it into three different scenarios: an upstream, a downstream and a jump-like upstream movement. The distributions of the wall heat flux, the quenching distance or the detachment of the maximum heat flux and the quenching point are strongly dependent on this movement. The highest heat fluxes appear mostly at the jump-like movement because the flame behaves locally like a head-on quenching flame.

Keywords: sidewall quenching; LES; premixed methane; flame–wall interaction; FGM

1. Introduction

The world's primary energy demand will be increasing about 30% in the next 20 years, and 75% of this demand will be provided by combustion [1,2]. Since resources are limited and combustion can cause pollution, it is necessary to better understand the fundamentals. One important field, in which research is required, is the flame–wall interaction (FWI). Due to technical improvements, e.g., of internal combustion engines or gas turbines, its relevance rises [3,4]. Here, the reaction zone of the flame gets close to cold walls and quenches. This reduces efficiencies and increases pollution formation [5]. The FWI can be subdivided into two canonical configurations, namely head on quenching (HOQ) and sidewall quenching (SWQ) [5]. In the latter case, the flame moves perpendicular along the wall and only the flame tip is affected. In case of HOQ, the flame moves parallel towards the wall and the whole flame quenches. The dimensionless quantities for the quenching distance and heat flux differ clearly in both cases. Because of the importance of the FWI, an SWQ configuration is considered in this work.

In terms of premixed conditions, SWQ has been widely studied theoretically, experimentally and numerically. Primarily, fuels like hydrogen and gaseous hydrocarbons, such as methane, ethylene, propane or butane, were used. The most simple fuel, but experimentally difficult to handle, is hydrogen. It was investigated by Cheng et al. [6] for an experimental turbulent boundary layer over a heated wall. They determined the limits of the reactions near the wall depending on the equivalence ratio. A more recent study regarding hydrogen combustion was conducted by Gruber et al. [7]. Within their three-dimensional direct numerical simulation (DNS), the authors used detailed chemistry (DC). Their data suggests that near-wall coherent turbulent structures are important for the wall heat flux and that the heat release rate at the wall is strongly controlled by exothermic radical recombination reactions. However, hydrogen behaves differently than hydrocarbon based fuels because it has a much higher consumption speed and it will not be investigated in this current work.

The effect on the quenching layer thickness of different hydrocarbons with variations in the equivalence ratio and different wall conditions was researched by Saffman [8]. The investigations of Lu et al. [9] and Ezekoye et al. [10] used different fuels and equivalence ratios as well, but in a constant volume chamber. They measured the unsteady heat transfer during SWQ. Boust et al. [11] developed a thermal formulation for single-wall quenching without any empirical coefficients. This formulation is based on experimental data and is applicable for lean and stoichiometric methane/air mixtures in a pressure range from 0.05–0.35 MPa. DNS studies combined with a single step mechanism, which mimics a lean methane flame with an inlet temperature of 600 K were carried out by Alshaalan and Rutland [12,13]. They investigated a V-shaped flame in a weakly turbulent Couette flow. Nevertheless, the number of simulations with a chemistry treatment beyond single step mechanisms for hydrocarbons is sparse and most turbulent premixed FWI investigations were focused on the HOQ configuration.

In the present numerical study, an SWQ burner configuration is investigated, which is based on the experimental design of Jainski et al. [14,15]. In the experimental setup, a premixed V-shaped flame interacts with a cooled wall under laminar and turbulent conditions. First numerical results for both cases have been already provided by Heinrich et al. [16,17] with a highly-resolved (large eddy simulation) LES with DNS-like spatial resolution. The turbulent and laminar cases were compared and analyzed with each other and the experiments. The global structure of the flame, the flame position, the temperature field and the velocities are given correctly by the methods used. Compared to other highly resolved simulations [7,12,13], which used simple configuration, this setup is more complex and the turbulent structures of the flow are not comparable with channel or pipe flows. In this work, a transient analysis of the flame–wall–turbulence interaction (FWTI) is conducted. This gives further insight into heat fluxes, quenching distances and quenching positions beyond the experimental findings. It will be shown that a strong dependency of the moving direction of the flame tip for characteristic properties exist. Moreover, it will be explained that the highest heat fluxes occur because the turbulent SWQ flame behaves locally like an HOQ flame.

The remainder of the paper is structured as follows: the numerical description, the chemistry treatment and the domain are given in Section 2. The transient analysis of the FWI is provided in Section 3. Finally, conclusions are given in Section 4.

2. Methods

2.1. Numerical Description

The simulation in this work has been conducted with the Finite Volume code FASTEST. It is a block-structured code that uses hexahedral, boundary fitted grids to represent complex geometries. The Massage Passing Interface (MPI) is utilized for the inter-processor communication. The incompressible Favre-averaged Navier–Stokes equations with a low-Mach number formulation for the variable density ρ are discretized with a second order spacial scheme [18] and the time integration is done by an explicit low storage three-stage Runge–Kutta scheme with second order accuracy. In each Runge–Kutta step, the pressure correction procedure is applied. The corresponding mass and momentum equations are the following:

$$\frac{\partial \overline{\rho}}{\partial t} + \frac{\partial}{\partial x_i}(\overline{\rho}\tilde{u}_i) = 0, \quad (1)$$

$$\frac{\partial}{\partial t}(\overline{\rho}\,\tilde{u}_j) + \frac{\partial}{\partial x_i}(\overline{\rho}\,\tilde{u}_i\tilde{u}_j) = \frac{\partial}{\partial x_i}(\overline{\tau}_{ij} - \overline{\rho}\tau_{sgs}) - \frac{\partial \overline{p}}{\partial x_j} + \overline{\rho}g_j. \quad (2)$$

The quantity u is the velocity, p is the pressure, τ_{ij} is the shear stress tensor and the operators ~ and ¯ represent the Favre-averaging and filtering operation, respectively. The sub grid fluxes of the momentum τ_{sgs} are accounted by eddy viscosity approach proposed by Smagorinsky [19]. The model coefficient is determined by the dynamic procedure proposed by Germano et al. [20] with a modification by Lilly [21]. For the discretization of the convective terms in the scalar transport equations, a total variation diminishing (TVD) limiter as proposed by Zhou et al. [22] is used.

2.2. Numerical Domain

Figure 1 displays the numerical domain used in this work, which is based on the experimental setup of Jainski et al. [14]. The dimensions are given on the left side of this figure. On the right side, an instantaneous velocity field of the center plane with the flame front is shown in the three-dimensional domain. The turbulent inlet flow and acceleration of the fluid due to the flame front is clearly visible.

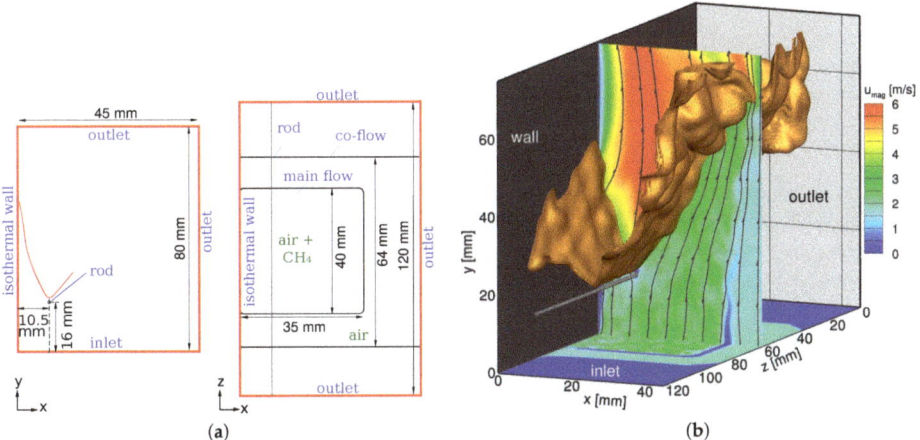

Figure 1. Numerical domain of the simulation. (**a**) dimensions of the domain; (**b**) snapshot of the 3D LES with velocity magnitude together with the flame front (orange iso-surface, $\dot{\omega}_{CO_2} = 60\,\text{kg}/(\text{m}^3\,\text{s})$).

Like in the experiments [14], the wall temperature is kept constant at $T_w = 350\,\text{K}$ and the rod temperature is set to 300 K. The numerical rod is fully resolved with an O-grid. To stabilize the stoichiometric methane main flow, it is surrounded by a pure air co-flow (see Figure 1a). The inlet temperatures of the main and co-flow are 300 K. The turbulent inlet flow was calculated in a pre-processing step by simulating the experimental burner geometry. Each time step, a turbulent velocity field is set at the inlet. How this was calculated will be explained in the next paragraph. At the outlets, convective boundary conditions are used. A block structured grid with approximately 21.5 million control volumes is applied. Distant from the wall, the spatial resolution ranges from 200 μm to 400 μm. Towards the wall, the mesh is refined to 90 μm. As discussed in [17] this grid leads to an direct numerical simulation (DNS)-like spatial resolution and no additional wall modeling is needed.

The complete validation and verification of the simulation of the non-reacting turbulent inlet flow can be found in Heinrich et al. [17]. Here, the basic idea will be pointed out. Therefore, the body of the experimental burner was simulated (Figure 2a) and the experimental mass flows for the main and co-flow were set as inlet conditions [14].

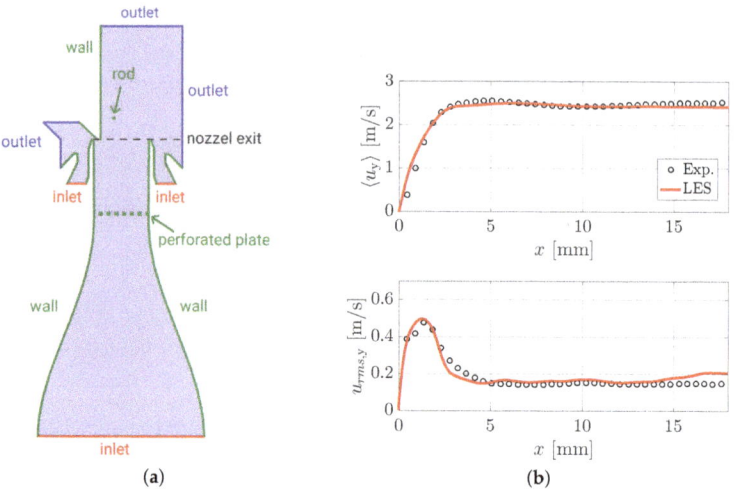

Figure 2. Characterization of the simulated burner geometry used for the turbulent inlet profiles. (a) sketch of the burner geometry. The boundary conditions are color-coded; (b) velocity profiles 11 mm after the nozzle exit.

The simulated geometry includes the main and co-flow, the perforated plate, the flame stabilizing rod and the wall. The height of the nozzle exit is marked in Figure 2a. The velocity fields were extracted along this area for each time step and stored in a database. Within the burner, the turbulence was generated by the perforated plate. After the nozzle exit, a boundary layer developed at the wall. The Reynolds number based on the nozzle exit is $Re = 5300$. The resulting turbulent intensity at the nozzle exit section is approximately 8%. The analyses of the flow field revealed that the structures are inhomogeneous. In Figure 2b, the time averaged velocity and its rms-value are shown for the main component 11 mm downstream the nozzle exit. Both quantities are in very good agreement with the experimental values, since the maximum difference between the profiles is less than 0.18 m/s.

2.3. Chemistry Treatment

Depending on the quantities of interest, a very detailed description of the near-wall chemistry may be required. However, the application of full reaction mechanisms, typically covering tens of species and hundreds of elementary reactions, is not possible in complex configurations. Even for the laboratory scale system considered here, the computational cost given by the resolution requirement, the number of transported scalars and their stiff coupling prohibit its application. Accordingly, a chemistry reduction approach by means of tabulation is employed. In the framework of the Flamelet Generated Manifolds (FGM) approach introduced by van Oijen and de Goey [23], pre-computed detailed chemistry solutions of premixed flames enter the chemistry database. By that, vital information is included such that important quantities like the flames' propagation speed or its spatial structure including all species can be exactly reproduced by means of the controlling variables. The table generation, its assessment in a FWI situation, and its treatment in turbulent flow are briefly outlined in the following.

2.3.1. Construction of the FGM Table

A three-dimensional chemistry table is used where the physical processes of this configuration being the reaction, the heat transfer to the wall and the mixing with the co-flowing air are accounted for by a reaction progress variable Y_{CO_2} (being the mass fraction of CO_2), the enthalpy h, and the mixture fraction Z, respectively. The table is illustrated in Figure 3b where the temperature and equivalence ratio have been chosen as the axis for illustrative purposes. As shown on the left, first a slice is generated by joining flamelets obtained with the GRI3.0 reaction mechanism [24,25] for different enthalpy levels. Hereby, region a) can be covered using different preheating temperatures while the burner stabilized method [26] is utilized in region b) where chemical reactions evolve under significant heat losses as present at the wall. Finally, in region c), the absence of activation energy suppresses most of the chemical activity and it mostly corresponds to a cooling of exhaust gases. Accordingly, the thermo-chemical states in this region can no longer be obtained by detailed chemistry simulations and Ketelheun et al. [27] approximated them by a thermo-dynamically consistent extrapolation. To cover mixing (i.e., different equivalence ratios), several of those slices are then combined to build the final three-dimensional table as given on the right of Figure 3b. Here, the source term $\dot{\omega}_{CO_2}$ has been added representing one of the quantities extracted during the lookup in the simulation.

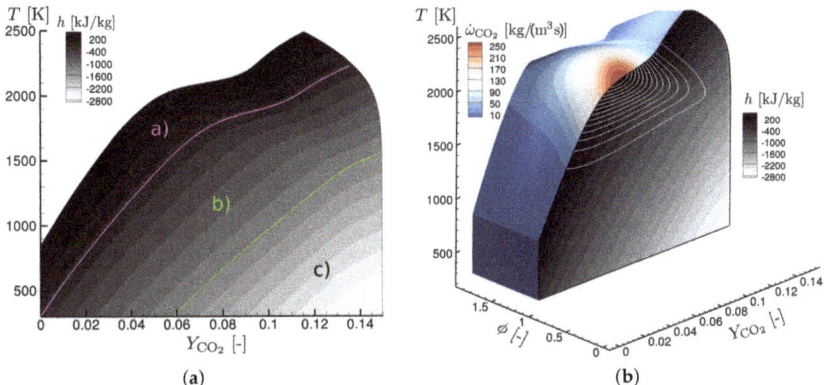

Figure 3. Representation of the table used. (a) cut at $\phi = 1$ with three flamelet types, a) freely propagating, b) burner stabilized, c) extrapolated. (b) 3D table with a cut at $\phi = 1$ and CO_2 source term $\dot{\omega}_{CO_2}$.

2.3.2. Evaluation of Simplifying Assumptions for FWI Application

The above-mentioned tabulation approach has been successfully applied to turbulent laboratory scale combustors [28–30] where the FWI is not a crucial part of the physics. With respect to the latter, some works have been conducted to evaluate the accuracy of FGM near cold walls. Specifically, Meier et al. [31] and Ganter et al. [32] performed comparisons to detailed chemistry simulations. It turned out that FGM can generally satisfactorily predict the flame attachment position and the transported major species as well as the temperature profiles. Pollutants such as the accumulation of carbon-monoxide near the wall could not be predicted by FGM (Ganter et al. [33]). They showed that this is also possible by advanced or at least more specific tabulation techniques, which, however, require further development. Based on the analysis, we consider that the FGM approach is sufficiently accurate for the quantities analyzed in this study.

Specific for this work and the assumptions made, Figure 4 shows the evaluation of the employed FGM in a generic, two-dimensional SWQ configuration. Detailed chemistry simulations represent the verification basis for the tabulation in general and also to justify diffusivity assumptions. Regarding the latter, within the table generation, the unity Lewis number assumption is employed, since it simplifies the mapping process due to having a constant enthalpy and mixture fraction throughout a flamelet. The under prediction of the flame speed caused by this assumption (about 30% for the relevant conditions) is then corrected afterwards by consistently altering the chemical sources and diffusivities which accelerates the flame according to classical flame theories [34] while preserving its structure [35].

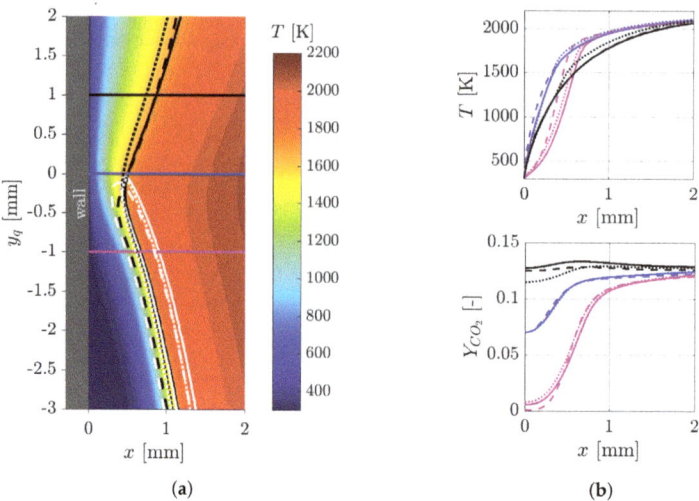

Figure 4. Comparison of different chemistry treatments: DC with mixture averaged diffusion (dashed lines), DC with $Le = 1$ assumption (solid lines) and FGM (dotted lines). (**a**) temperature field of the DC simulation. The black lines represent iso-lines of $T = 1700$ K and the white lines of $\dot{\omega}_{CO_2} = 60$ kg/(m^2 s); (**b**) the colors of the lines represent the extracted position from (**a**). The top figure shows the corresponding temperature lines and the bottom figure the mass faction of CO_2.

First, Figure 4a shows the temperature field in the region around the quenching point for orientation. Added to this graphs are lines marking the flame by means of an isothermal as well as the chemical source term obtained by three different simulations. The first (dashed line) is a detailed chemistry simulation employing individual diffusivities for all species and can be considered as the most accurate physical description. The Second (solid line) is also a detailed chemistry simulation but with the $Le = 1$ assumption. Finally, the dotted line is the FGM solution. Due to the above-mentioned correction of the flame speed, the flame–wall attachment points and flame angles are in close agreement. Slightly upstream of the quenching point (coordinate zero) the individual diffusivities cause the flame to reach slightly closer to the wall while the FGM solution is close to the $Le = 1$ detailed chemistry indicating that these deviations are due to the remaining transport simplifications. Downstream of the quenching point, the isothermal of the detailed chemistry simulations then approach each other while the FGM slightly depart from them. Accordingly, and as also analyzed by Ganter et al. [33], this is due to tabulation assumption regarding diffusive fluxes orthogonal to the controlling variables. As mentioned, correcting this would require an immensely increased tabulation effort being currently not yet applicable to this configuration. Considering the rather small errors this seems acceptable. A quantification of the temperature and species evolution is furthermore provided on the right of Figure 4 where profiles are shown extracted along the horizontal lines added on to the left of Figure 4. Here, the blue, magenta and black lines represent the states at the quenching point and one millimeter up- and downstream of it. All of these positions being in the region of intense FWI confirm a very good approximation by the FGM approach for the purpose of this work.

2.3.3. Treatment in the Turbulent FWI Flow

As mentioned above, the thermo-chemical state is described with three transported quantities, the mixture fraction Z, the reaction progress variable Y_{pv} and the enthalpy h:

$$\frac{\partial \bar{\rho}\tilde{Z}}{\partial t} + \frac{\partial}{\partial x_j}\left(\bar{\rho}\tilde{u}_j\tilde{Z}\right) = \frac{\partial}{\partial x_j}\left(\left(\mathcal{FE}\frac{\overline{\mu}}{Sc} + (1-\Omega)\frac{\mu_t}{Sc_t}\right)\frac{\partial \tilde{Z}}{\partial x_j}\right), \tag{3}$$

$$\frac{\partial \bar{\rho}\tilde{Y}_{pv}}{\partial t} + \frac{\partial}{\partial x_j}\left(\bar{\rho}\tilde{u}_j\tilde{Y}_{pv}\right) = \frac{\partial}{\partial x_j}\left(\left(\mathcal{FE}\frac{\overline{\mu}}{Sc} + (1-\Omega)\frac{\mu_t}{Sc_t}\right)\frac{\partial \tilde{Y}_{pv}}{\partial x_j}\right) + \frac{\mathcal{E}}{\mathcal{F}}\dot{\omega}_{pv}, \tag{4}$$

$$\frac{\partial \bar{\rho}\tilde{h}}{\partial t} + \frac{\partial}{\partial x_j}\left(\bar{\rho}\tilde{u}_j\tilde{h}\right) = \frac{\partial}{\partial x_j}\left(\left(\mathcal{FE}\frac{\overline{\mu}}{Sc} + (1-\Omega)\frac{\mu_t}{Sc_t}\right)\frac{\partial \tilde{h}}{\partial x_j}\right), \tag{5}$$

where μ is dynamic viscosity, Sc the Schmidt number, $\mathcal{F}$ the thickening factor, $\mathcal{E}$ the efficiency function, Ω the flame sensor and $\dot{\omega}_{pv}$ the reaction source term. To capture the flame–turbulence interaction correctly, the artificially thickened flame (ATF) approach is used. In this context, the flame has been thickened with $\mathcal{F}$ to resolve its structure on the LES grid. Therefore, the transformation of Butler and O'Rouke [36] is used, where the flame speed is preserved. A dynamic version of the procedure based on a sensor Ω is used in this work [37,38]:

$$\mathcal{F} = 1 + (\mathcal{F}_{max} - 1)\Omega, \tag{6}$$

$$\Omega = 16\left(Y_{n,pv}(1 - Y_{n,pv})\right)^2, \tag{7}$$

where $\mathcal{F}_{max}$ is the grid depending maximum thickening factor and $Y_{n,pv}$ is the normalized progress variable:

$$Y_{n,pv} = \frac{Y_{pv} - Y_{pv,min}}{Y_{pv,max} - Y_{pv,min}}. \tag{8}$$

Due to the small thickening factor $\mathcal{F}_{max}$ of 1.4 in this work, the flame turbulence interaction is only slightly altered. Since ATF is used for the flame–turbulence interaction, the turbulent diffusion is switched of in the region of the flame. As explained in [16], the influence of the dynamic thickening on relevant quantities, like wall heat flux, the flame position and quenching point determination is negligible because the thickening can be retransformed. In general, the interaction with the turbulence is reduced if the flame is artificially thickened. To compensate these effects, Colin [35] used a so-called efficiency function $\mathcal{E}$. The formulation of the implemented efficiency function $\mathcal{E}$ can be found in Charlette et al. [39]. Within this work, $\mathcal{E}$ is nearly negligible because the grid resolution is fine enough and $\mathcal{F}_{max}$ is relatively small. Following the classical turbulent premixed combustion diagram from Peters [40], this setup operates in the wrinkled flamelet regime, meaning that the inner flame structure is close to a laminar flame and the thin flame front is wrinkled by the turbulent motion due to the fact that the flamelet concept is valid [40]. Additionally, the effects of curvature and stretch on the mass burning rate have been investigated for the thin reaction zone regime [41] and the corrugated flamelet regime [42]. The resulting errors have been found to be within 5% when using a single progress variable as done in this work. Therefore, FGM can be applied for the considered configuration.

2.4. Determination of the Quenching Point

Following Heinrich et al. [16], the quenching point y_{qp} is determined by tracking the maximum value of the CO_2 source term, which defines the flame front. Due to the enthalpy losses at the wall, this source term decreases and if it drops under a certain threshold the flame is quenched. The threshold is defined as the half of the maximum CO_2 source term along the flame front. This procedure is analogous to the experimental determination where the OH gradient was used. For this FGM approach, using the CO_2 source term is validated against detailed chemistry simulations and it appears to be the most reliable method [16].

2.5. Comparison with Experimental Data

As mentioned before, the models used lead to good results compared to the experimental laminar and turbulent SWQ burner. This was shown before by Heinrich et al. [16,17]. Within these references, the comparisons of the laminar and turbulent for the non-reacting and reacting configurations can be found. For the analyses conducted in Section 3, the velocity, flame position and quenching points are of importance. For the sake of completeness, these quantities are shown again in Figure 5.

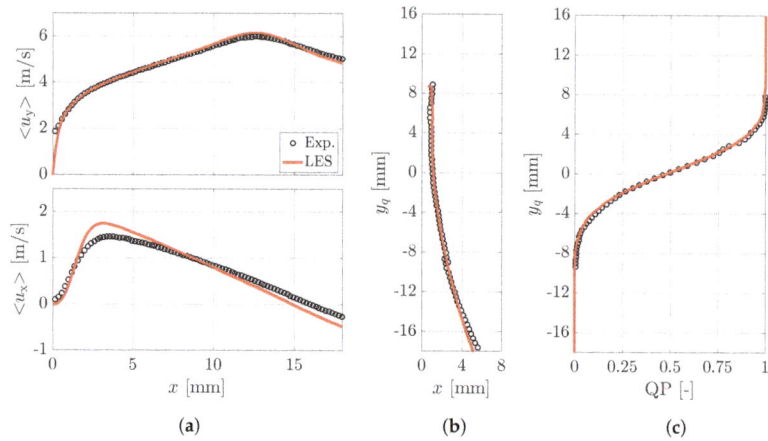

Figure 5. Comparisons of the turbulent data with experiments at $z = 60$ mm in the conditioned coordinated system. (a) time averaged velocity profiles 5 mm upstream the quenching point; (b) averaged flame position; (c) quenching probability.

Firstly, a conditioned coordinated system has to be introduced (x_q, y_q) with respect to local averaged quenching point is introduced. It is centered at the wall with $x_q = 0$ mm and $y_q = 0$ at the averaged axial quenching point (exemplary marked with the black solid circles in Figure 6). The experimental data is taken from Jainski et al., 2016 [14]. The velocity profiles 5 mm before the time averaged quenching point are shown in Figure 5a. Both directions recover the experimental velocity very well. The difference in the main flow direction is less than 0.17 m/s, except for the point at the wall. Clearly, at the wall, an experimental inaccuracy occurred because the the velocity is not zero. In the wall normal direction, the maximum difference is slightly higher (0.3 m/s). The averaged flame positions (Figure 5b) are in excellent agreement. The maximum deviation is in order of the grid size, 0.2 mm. Figure 5c depicts a statistical quantity, the quenching probability. If it is zero, the flame is always burning and, if it is one, the flame is always quenched. Both profiles match perfectly. It shows that quenching events takes place in the region -8 mm $< y_{qp} < 8$ mm.

2.6. Dimensionless Parameters

Two dimensionless parameters used for the analyses have to be explained. The Péclet number Pe describes the ratio of the wall normal distance of the flame and a typical flame thickness:

$$\text{Pe} = \frac{x}{\delta}.$$

The diffusive flame thickness is used as reference to define the Péclet number at quenching Pe_q [5]:

$$\text{Pe}_q = \frac{x_q}{\rho_u c_{p_u} S_L / \lambda_u} \frac{1}{f_{\text{atf}}}, \tag{9}$$

where ρ is the density, c_p is the specific heat capacity at constant pressure, S_L is the laminar flame speed and λ refers to the thermal conductivity of the gas. The occurrence of f_{atf} is due to the ATF transformation as detailed in Heinrich et al., 2017 [16]. The subscript q refers to values at the quenching point and the subscripts u and b refer to unburnt and burnt states. The wall heat flux $\Phi_{w,q}$ is made dimensionless with the flame power $P_Q = \rho_u S_L c_{p_u}(T_b - T_u)$ [5]:

$$F_Q = \frac{\Phi_{w,q} \cdot f_{\text{atf}}}{P_Q}, \tag{10}$$

where T is its local temperature. The applied parameters are given in Table 1.

Table 1. Characteristic values from the CHEM1D calculation for a premixed freely propagating flamelet with 300 K inlet temperature, stoichiometric conditions and mixture averaged diffusion.

T_u	T_b	S_L	c_{p_u}	ρ_u	λ_u
300 K	2286 K	37.1 cm/s	1077 J/(K kg)	1.123 kg/m^3	0.0283 W/(m K)

3. Results

To provide a proper understanding, this section is divided into three parts. First, a mostly descriptive insight into the transient FWI is given to provide a basic understanding on the typical physical behavior and the associated quantities like the heat flux. Therefore, first, Section 3.1 illustrates the basic evolution of the FWI depending on the type of the quenching point displacement by means of individual, representative events. The observations are then shown in Section 3.2 to also hold on a statistical basis where further observations are provided. Finally, Section 3.3 outlines the root causes by considering the full interaction of the flame, the wall and the turbulent field (FWTI). It is exemplified by following the mechanisms leading to very large heat fluxes associated with a departure from the SWQ towards the HOQ regime.

The averaged quenching points along the wall are shown as a red solid line in Figure 6 together with all quenching points from one snapshot (black solid line). These lines are called quenching lines. As it can be seen, the averaged quenching line is not constant and, for an instantaneous moment, the flame can quench far away from this line. In this study, three independent planes will be used for the transient analysis of the FWI, at the center ($z = 60$ mm), at $z = 40$ mm and at $z = 80$ mm (blue dashed lines). These planes are uncorrelated for the quenching point statistics since the correlation-coefficient drops under zero at a length of 10 mm. Since the averaged quenching line is not constant in the z-direction, the conditioned coordinate system (x_q, y_q) introduced before is used. With this transformation, the three extracted lines can be combined to one dataset.

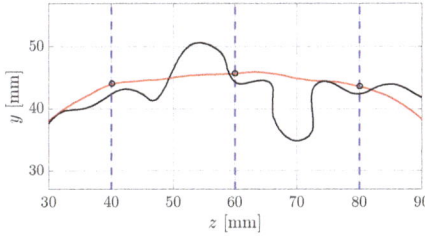

Figure 6. Time averaged quenching points along the wall (red solid line) with one instantaneous quenching line (black dashed-dotted line) and the three extracted planes marked (blue dashed lines).

3.1. Description of the Quenching Point Movement

As pointed out by Heinrich et al. [17], the quenching positions, heat fluxes and wall angles are widely distributed. In this work, it will be shown that most of these statistics depend on the way the flame tip moves. The movement of the flame tip is tracked via the temporal quenching point position. A sequence from the center plane is shown in Figure 7, where the downstream and upstream movement are colored in green and red, respectively.

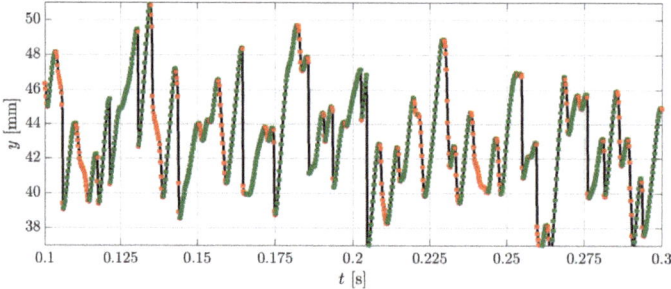

Figure 7. Temporal section of the axial quenching point location with colored downstream and upstream movement (green and red). The interval between the points is 0.1 ms.

Clearly, this movement seems to be chaotic and is not equally distributed. In 71% of the events, the flame tip is moving in the downstream and in 27% in the upstream direction. Consequently, the averaged upstream velocity of the flame tip must be faster than its averaged downstream velocity. The remaining 2% are a jump-like upstream movement, with multiple quenching points along the wall. This leads to the following categorization:

(A) A downstream movement,
(B) A moderate upstream movement,
(C) A jump-like upstream movement with multiple quenching points.

The following paragraphs together with the Figures 8–10 describe the scenarios in more detail. The earliest time step is shown at the top and the latest on the bottom. The individual time steps are organized as follows (from the left to the right):

- two-dimensional contour plot of the wall heat flux together with the quenching line (red line), the position of the extracted profiles ($z = 0$ mm, white line) and the moving direction of the flame tip (white arrow)
- Φ_w along the wall at $z = 0$ mm, including the position of the maximum wall heat flux point $y(\Phi_w)$ (blue dashed line) and y_{qp} (red dashed line),
- a 2D slice (x-y-plane) with temperature isolines at $z = 0$ mm with the flame position (black line) overlaid with the velocity vectors and $y(\Phi_w)$ (blue dashed line) and y_{qp} (red dashed line),
- the temperature field including the magnitude of the velocity and its vector for two positions at $z = 0$ mm (one at y_{qp}, one at $y_{qp} - 4\delta_T$, where δ_T is the thermal thickness of the flame).

The explanations and statistics for scenarios are provided in the next sections.

(Scenario A)

The moderate downstream movement of the flame tip appears in nearly 71% of the cases, which makes it the most frequent event. How the flame and the wall heat flux are connected is explained in conjunction with Figure 8.

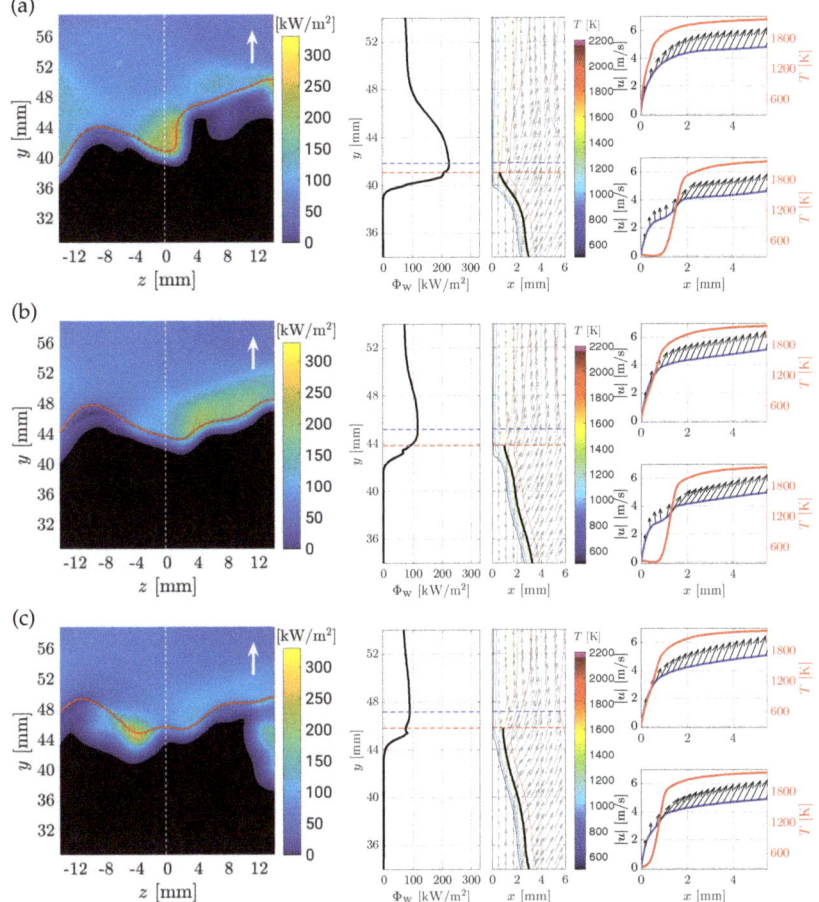

Figure 8. (Scenario A): Consecutive snapshots (**a**–**c**) showing the time sequence of the downstream movement of the quenching point. From the left to the right are the wall heat flux with all quenching points shown, then Φ_w along the wall at $z = 0$ mm, including the position of the maximum wall heat flux point $y(\Phi_w)$ (blue dashed line) and y_{qp} (red dashed line). The next graph displays a 2D slice (x-y-plane) with T-isolines at $z = 0$ mm with the flame position (black line) overlaid with the velocity vectors and $y(\Phi_w)$ (blue dashed line) and y_{qp} (red dashed line). The last graph shows the temperature field including the magnitude of the velocity and its vector for two positions at $z = 0$ mm (one at y_{qp}, one at $y_{qp} - 4\delta_T$, where δ_T is the thermal thickness of the flame).

The quenching point y_{qp} and the location of the maximum wall heat flux $F_{Q,max}$ move together in an axial direction, while the heat flux constantly decreases. As visible, $F_{Q,max}$ is downstream of the quenching point. The difference between both locations can be up to 12 thermal flame thicknesses. Within this movement and upstream of y_{qp}, mostly cold fresh gases flow parallel to the wall. This can be seen at the temperature field and at the extracted profiles. As visible on the right graphs, directly at the wall and 2 mm before the quenching point, the fluid is not heated by the flame. Compared to the laminar case, the heat fluxes are rather low and the quenching distances are large ($\Phi_{w,l} = 200$ kW/m^2, $Pe_{q,l} = 7.3$).

(Scenario B)

The next scenario is depicted in Figure 9.

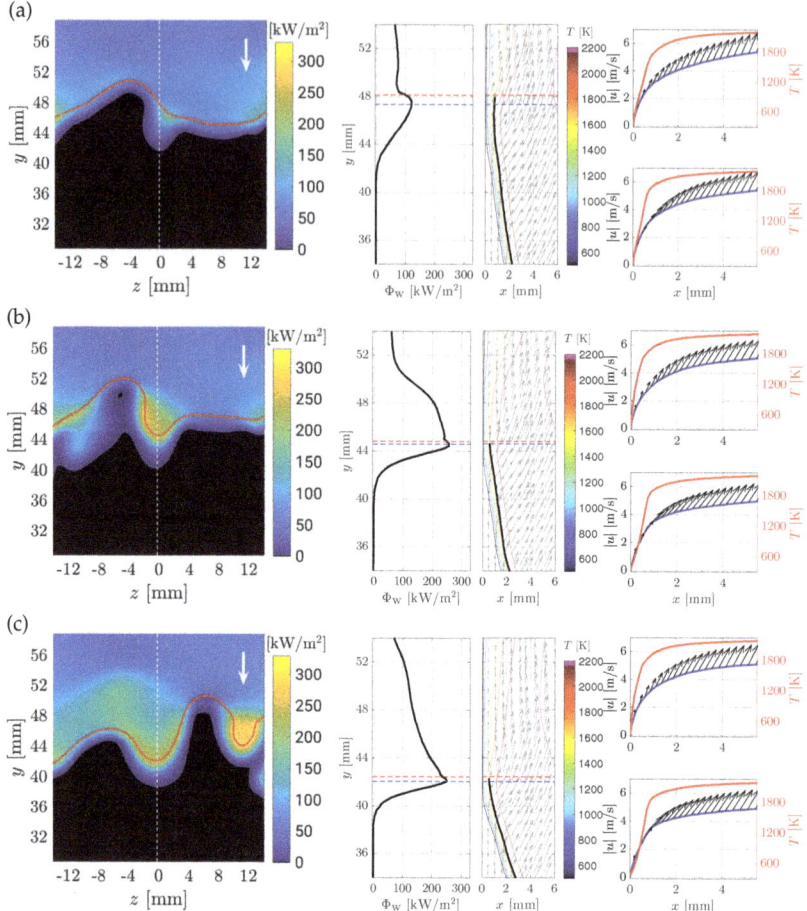

Figure 9. (Scenario B): Time sequence of the upstream movement of the quenching point. The explanation of the figure can be found in Figure 8.

This moderate upstream movement with increasing maximum wall heat flux occurs in 27% of the events. The maximum heat fluxes and the quenching point are closely together, like in the laminar case, while the heat flux is mostly higher than in (scenario A) or than the laminar reference value $\Phi_{w,l} = 200$ kW/m^2. Within this kind of movement, only high heat fluxes and low Pe_q can be found. In contrast to (scenario A), no explicit temporal behavior of $\Phi_{w,max}$ can be distinguished; within this movement, it can increase or decrease. Additionally, it can be seen from the temperature profiles on the bottom at right side of Figure 9 that the fluid upstream of the quenching position is heated by the flame.

(Scenario C)

The last scenario is found only in 2% of the cases. However, it cannot be neglected because only the strongest heat fluxes arise there. It is also an upstream movement like (scenario B), but it appears to look like a jump of the quenching point. Figure 10 will be used to explain the behavior. At the beginning (a), the flame front is nearly parallel to the wall and moves downstream like in (scenario A), but the wall heat flux upstream the quenching point increases at a large region. At the same moment, the flame gets weakened where the heat flux rises and the enthalpy near the wall decreases in the same area (not shown here). A widespread quenched region emerges (Figure 10b) and the heat flux reaches extreme values (up to $F_{Q,max}/F_{Q,max,l} = 1.7$). In this moment, three quenching points occur in the center plane and the moving direction of the flame tip flips. Since the quenching point most upstream is always used as the reference point, it seems like it jumps upstream. As it can be seen in the left graph, in the next moment (c), both quenching areas will merge together and only one quenching point will remain. During this process, the occurring values of $F_{Q,max}$ are only the highest values. Afterwards, the flame tip moves downstream again with a lower heat flux, being the transition to (scenario A).

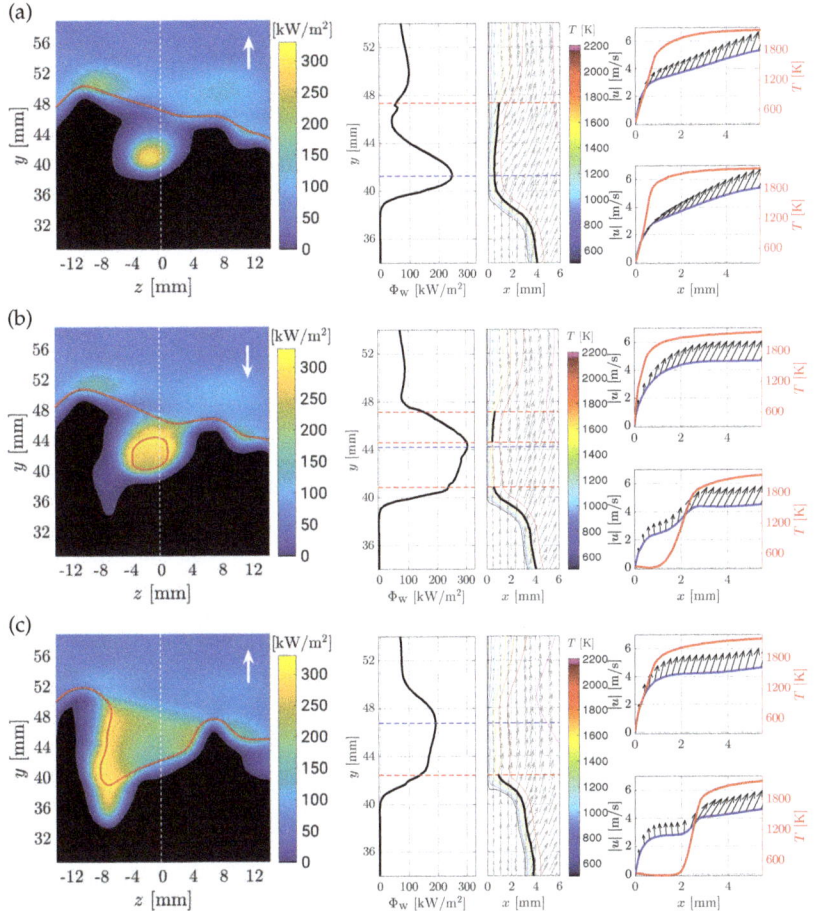

Figure 10. (Scenario C): time sequence of the upstream jump of the quenching point. The explanation of the figure can be found in Figure 8.

3.2. Statistics of the Axial Quenching Position and the Maximum Wall Heat Flux

As shown in Figure 11a, the locations of the maximum wall heat flux $y(F_{Q,max})$ (blue solid line) and the quenching point y_{qp} (red dashed line) are at the same position in a laminar SWQ flame. The black crosses correspond to the dimensionless wall heat flux profile (top x-axis) and the magenta line displays the iso-line $\dot{\omega}_{CO_2} = 60\,\mathrm{kg/(m^2 s)}$, which envelops the flame front. These profiles are shown in Figure 11b for one instantaneous moment of the turbulent SWQ configuration. As already observed by Heinrich et al. [17], in a turbulent setup, y_{qp} and $y(F_{Q,\max})$ do not always coincide. This phenomenon is visualized with the black arrow in Figure 11b. The quantity $\Delta(y(F_{Q,\max}), y_{qp}) = y(F_{Q,\max}) - y_{qp}$ expresses the distance between the axial locations. A positive distance implies that the maximum heat flux is farther downstream. $F_{Q,\max}$ and y_{qp} are spacial quantities that change during time.

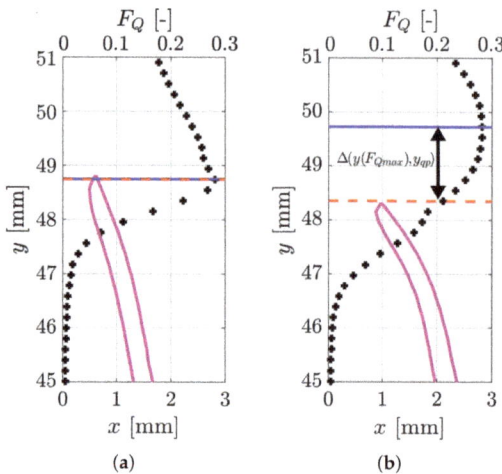

Figure 11. Flame position with marked quenching point (y_{qp}, red dashed line) together with the dimensionless wall heat flux (black crosses, top x-axis) and its maximum marked ($y(F_{Q,\max})$, blue solid line). The magenta line corresponds to $\dot{\omega}_{CO_2} = 60\,\mathrm{kg/(m^2\ s)}$, which envelops the flame front. (a) laminar; (b) turbulent: the black arrow shows the difference between y_{qp} and $y(F_{Q,\max})$.

First, the temporal distribution of $y(F_{Q,\max})$ and y_{qp} are analyzed in Figure 12a,b.

For both moving directions, y_{qp} is centered at zero and normal distributed (magenta dashed lines). However, the mean value of $y(F_{Q,\max})$ is zero for the upstream movement with equal number of counts the positive and negative side but with larger positive range. In contrast to that, the downstream direction has a mean value around 2 mm and more events occur on the positive side. The moving direction does not affect y_{qp}, but $y(F_{Q,\max})$ because it is shifted for the downstream movement, which could be seen in Figure 8 in the graphs in the middle. The same observation can be made by looking at Figure 12c,d, where $\Delta(y(F_{Q,\max}), y_{qp})$ is shown for all movements. In the following, most of the figures include similar histograms. A solid black line represents the mean value of the shown data and the dashed black the laminar reference. A cumulative histogram in light blue is overlaid. As it can be seen, within the downstream movement, for more than 60% of all events, the maximum wall heat fluxes are farther downstream. As a result, the mean difference is 1.8 mm. Like expected from the opposite direction (Figure 12b) and also visible in Figure 12d, $\Delta(y(F_{Q,\max}), y_{qp})$ is very small (80% are in the range $|\Delta(y(F_{Q,\max}), y_{qp})| \leq 1$ mm). Only a few outliers occur. The jump-like movement was not considered here because it is a very fast process whereby the determination of the local detachment is error-prone. Figure 12 showed that the shift between y_{qp} and $y(F_{Q,\max})$ is not constant and is mainly caused by the downstream movement of the flame tip.

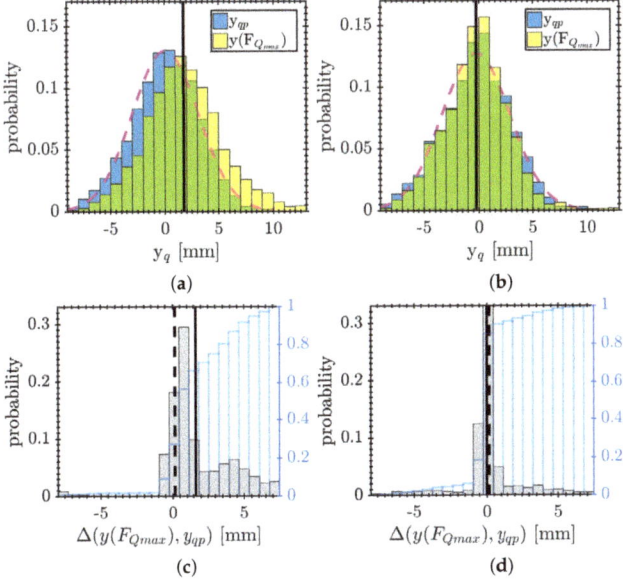

Figure 12. Relation between y_{qp} and $y(F_{Q,max})$ in dependency of the flame tip movement. (**a**,**b**): y_{qp} with corresponding normal distribution (magenta dashed line) and $y(F_{Q,max})$ with corresponding mean value (black solid line). (**c**,**d**): Mean values (black solid lines) and laminar reference (black dashed line). (**a**) Downstream. (**b**) Upstream. (**c**) Downstream. (**d**) Upstream.

Figure 13 displays the probability distribution of the maximum dimensionless wall heat flux. All three scenarios show a different behavior. If the flame tip moves downstream, only low values appear: nearly 90% are lower than the laminar reference. On the contrary, in the upstream scenario, 80% are higher than the laminar value. Accordingly, the mean values for these two cases are 0.155 and 0.27, respectively. The jump-like movement is most extreme; here, in 99% of the events, $F_{Q,max}$ is higher than the laminar scenario; accordingly, mostly the highest observed values appear. The distribution of the mean value of $F_{Q,max}$ depends on the movement of the flame tip; this was also shown in the representative snapshots in Figures 8–10.

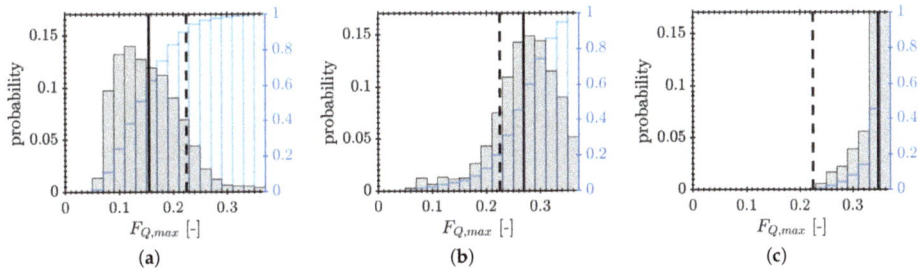

Figure 13. Dependency of $F_{Q,max}$ from the flame tip movement with its mean value (black solid line) and the laminar reference value (black dashed line). (**a**) downstream; (**b**) upstream; (**c**) jump-like.

Another quantity, which was not discussed before, is the dimensionless heat flux at the quenching point $F_{Q,q}$. Since it has a similar qualitative behavior to $F_{Q,max}$, its distribution will not be shown here. The difference is that the values for $F_{Q,q}$ are a bit lower and the maximum values only appear at the upstream and jump-like movement. At the upstream movement, nearly 70% of the events are higher than the laminar reference with a mean value of $F_{Q,q} = 0.255$. The opposite applies for the downstream direction, almost 95% are smaller, with a mean value of 0.12. Both distributions ($F_{Q,q}$ and $F_{Q,max}$) showed that high heat fluxes appear mostly while the flame tip moves upstream and the lowest values if it moves downstream.

As known from the literature (e.g., [5,17]), there is a negative correlation between Pe_q and $F_{Q,q}$. The cross-correlation between Pe_q and $F_{Q,q}$ is very strong, the correlation coefficients are -0.87 and -0.97 for the downstream and the upstream movement, respectively. With this in mind, the distribution of Pe_q Figure 14 is not unexpected.

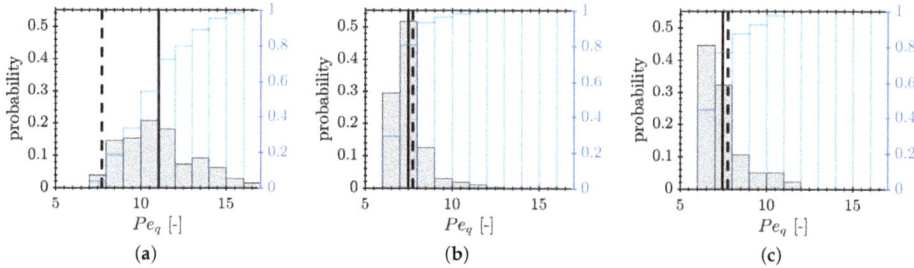

Figure 14. Dependency of Pe_q from the flame tip movement with its mean value (black solid line) and the laminar reference value (black dashed line). (**a**) downstream; (**b**) upstream; (**c**) jump-like.

For the downstream direction, almost 90% of the events are higher than the laminar reference and the mean value is about $Pe_q = 12$. This means that the flame is much farther away from the wall. The smallest possible wall distances do not occur. In contrast to that, only small distances occur for the upstream movement and its mean value $Pe_q = 7.5$ is slightly below the laminar case. Pe_q has a narrow distribution centered around its mean value with over 50% of the events in its proximity. The distribution for the jump-like movement is qualitatively similar to the upstream direction and the mean value $Pe_q = 7.5$ is slightly lower than the laminar value. Certainly, the most events occur around $Pe_q = 6.5$.

3.3. Physical Mechanism Governing the FWTI

Figure 14 displayed an obvious dependency of Pe_q from the moving direction and, with its help, the occurring wall heat fluxes can be explained, since the flame can get close to the wall and a strong temperature gradient occurs. The resulting heat fluxes are very high or the opposite can occur, the flame is far away and the heat fluxes are much smaller.

The behavior for the downstream and upstream distributions can be explained with the temperature field. For that reason, time averaged temperature fields with respect to the conditioned coordinate system (x_q, y_q) are shown in Figure 15b,c.

Figure 15a shows the laminar temperature field for comparing purpose. Compared to the laminar case, in the downstream movement, more cold unburnt fluid is enclosed between the flame and the wall, which pushes the flame farther away from the wall. The consequence of this was shown before, the resulting heat flux is smaller. While the flame tip moves in an upstream direction, the temperature field is very close to the laminar case as already shown in Figure 14b. In addition, the flame shapes are different, depending on the movement. In case of the downstream movement, the curvature is stronger.

For a laminar SWQ flame, Φ_w is always smaller than in the corresponding HOQ flame [5], but, as shown before, in the turbulent case, the wall heat flux can clearly exceed the laminar reference values up to values of the HOQ case or even higher. Due to that, the turbulent case will be compared qualitatively and quantitatively with an HOQ flame. Furthermore, a one-dimensional HOQ flame is calculated with FGM for this purpose. It has the same mixture properties as the main flow of the SWQ configuration and the same wall temperature. The one-dimensional grid used has the same resolution as the wall normal direction of the 3D SWQ case. As seen before in Figure 13, the jump-like movement has only the highest $F_{Q,max}$ values. This can be explained by analyzing this movement in a different conditioned coordinate system. Therefore, the location of the maximum wall heat flux $F_{Q,max}$ defines the zero on the y-axis. Figure 16 displays the time averaged temperature fields for all jump-like movements and the laminar reference.

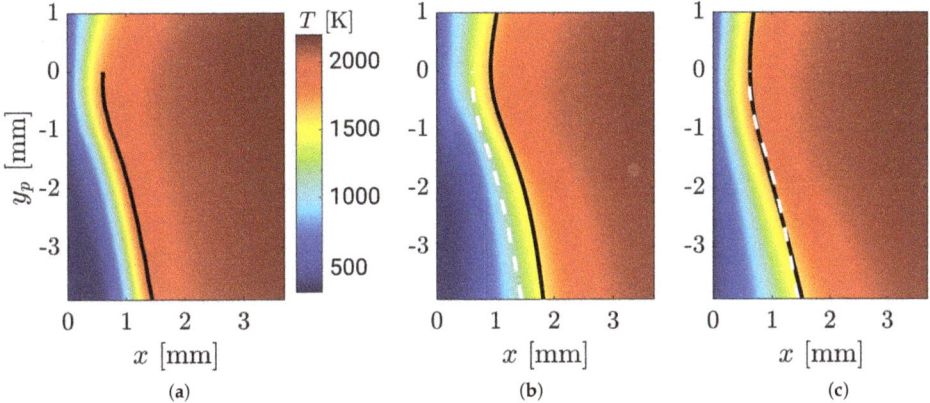

Figure 15. Dependency of the temperature field from the flame tip movement within the conditioned coordinate system together with the flame (black line). The white dashed line in (**b**,**c**) corresponds to the laminar reference. (**a**) laminar case; (**b**) downstream; (**c**) upstream.

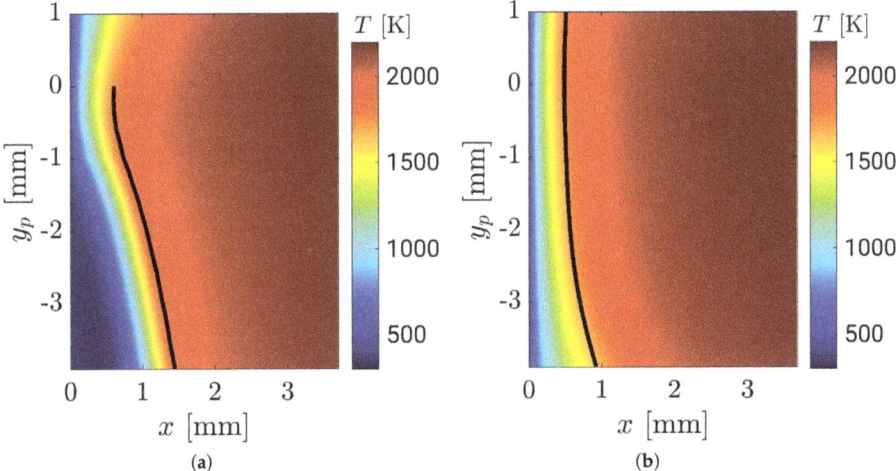

Figure 16. Temperature field in the conditioned coordinate system, where $y(F_{Q,max})$ marks zero. (**a**) laminar case; (**b**) jump-like.

As it can be seen on the right side, for the jump-like movement, the flame is nearly parallel to the wall, like in an HOQ manner. In addition, the temperature field is much more compressed in the direction of the wall, which causes the high heat fluxes. For a one-dimensional HOQ flame, the velocities near the wall are nearly zero at the moment the flame quenches. The same can be seen for the jump-like scenario in Figure 17.

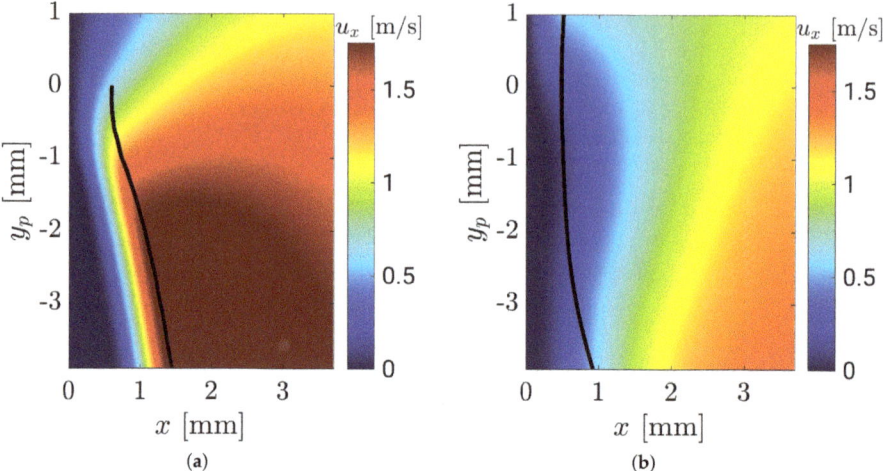

Figure 17. Wall normal velocity in the conditioned coordinate system, where $y(F_{Q,max})$ marks zero. (**a**) laminar case; (**b**) jump-like.

Compared to the laminar case, where the flame is clearly an SWQ flame, the wall normal velocities are much smaller. Due to that, the flame can burn much closer to the wall because the convection away from it is much weaker.

Since the turbulent SWQ flame front is transient and three-dimensional, a point of interest has to be tracked in time and space. How it is done will be explained with Figures 18–20.

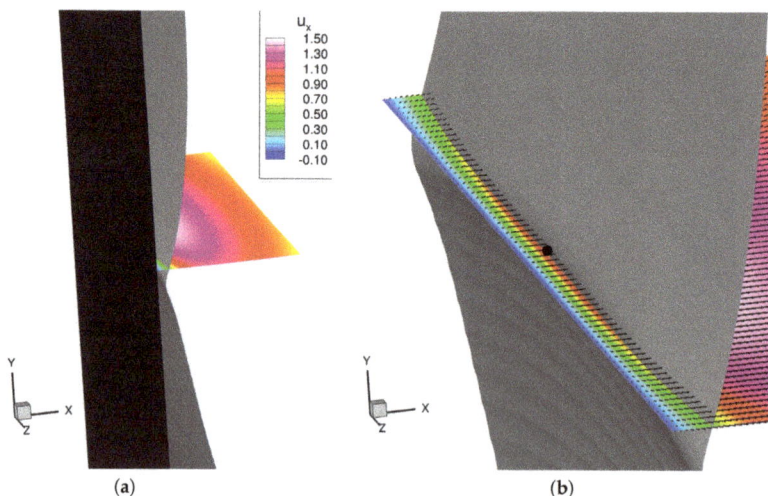

Figure 18. 3D flame front for the laminar case. (**a**) 3D flame front with wall (black surface); (**b**) zoom with monitored point (black dot).

A section of the 3D laminar flame is shown in Figure 18. On the left side, the wall is depicted for orientation purpose. The figure includes a temperature iso-surface of $T = 1750\,\mathrm{K}$ and a slice of the wall normal velocity component at the height of the maximum wall heat flux at the center of the burner. On the right side, a close-up view of the same section without the wall is displayed and the monitored quenching point is marked (black dot). Additionally, the momentum vectors are projected onto the slice. As it can be seen, the momentum vectors are all aligned and are pointed perpendicular away from the wall. The fluid is accelerated away from the wall by the flame without any redirection. Due to this convection, the flame cannot get closer to the wall.

The temporal evolution of the tracked point in turbulent case is shown in Figures 19 and 20, (black dots). It follows the path of a point on the flame front, which leads to the maximum wall heat flux at $t = 0\,\mathrm{ms}$.

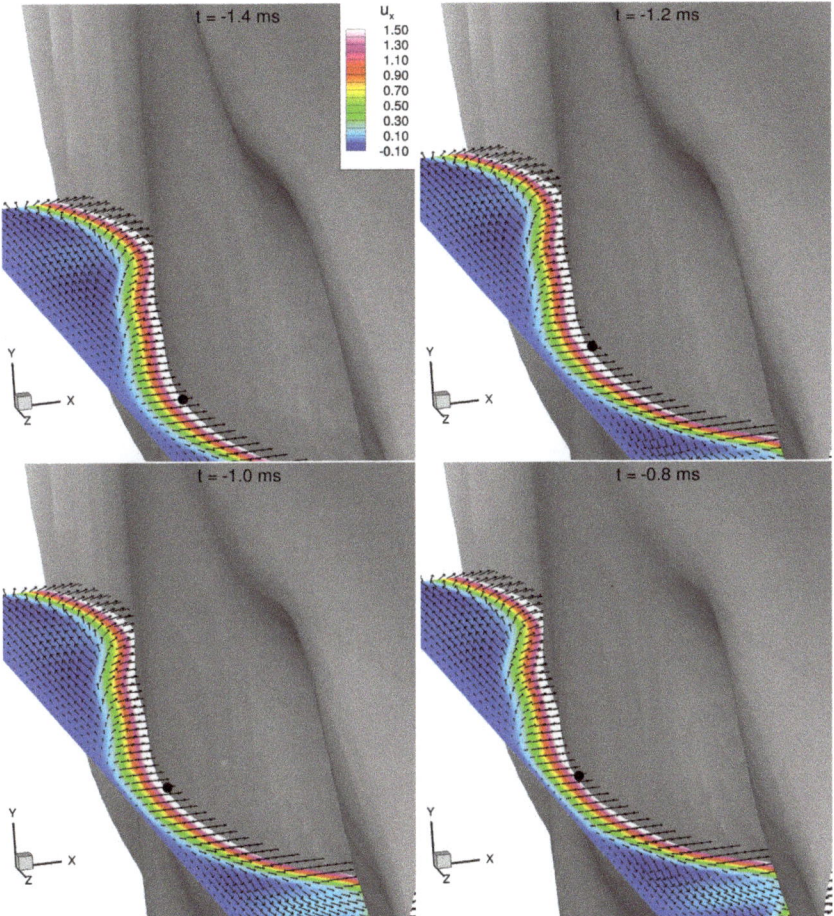

Figure 19. Temporal evolution of the 3D flame front near the wall, part 1.

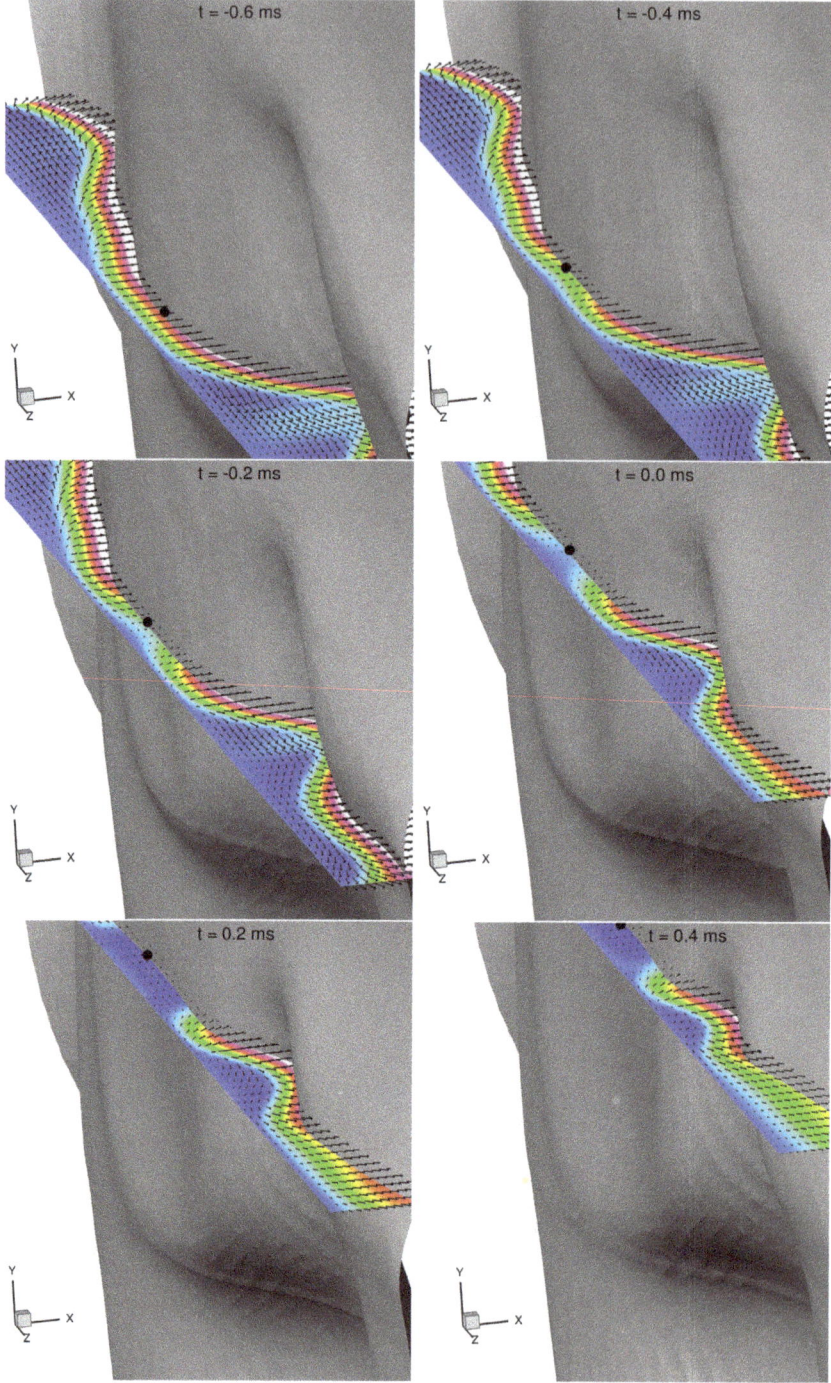

Figure 20. Temporal evolution of the 3D flame front near the wall, part 2.

As visible in the first sub-figure (top, left), the flame front is shaped convexly towards the unburnt gases. In contrast to the laminar case, the momentum vectors, which are not in the proximity of the tracked point, are directed away from the point into positive and the negative z-direction. During this progress, the wall normal velocity decreases and the convex flame shape gets flattened. Due to the convection of the main flow in the axial direction, the tracked point moves upstream while it stays more or less at the same z-position (which does not look like it is due to the chosen angle of view). By following this point, the graphs in Figures 21 and 22 were created.

The temporal evolution of the tracked point from the turbulent SWQ setup, an HOQ flame and the laminar reference case are shown in Figure 21. It displays from the left to the right: the wall heat flux, the normal velocity and the position of the flame.

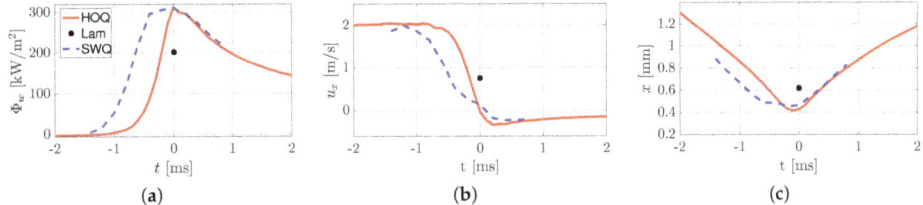

Figure 21. Temporal evolution of typical flame quantities for different flame types. (a) wall heat flux; (b) wall normal velocity at $T = 1750$ K; (c) position of the flame at $T = 1750$ K.

The time point $t = 0$ ms marks the moment with the highest wall heat flux. The position and the velocity of the flame front are extracted, where the temperature is equal to 1750 K. This location coincides with the position of the maximum source term of the flame and therefore it is associated with the center of the flame.

At first, the 1D HOQ case will be illustrated. At the beginning, the heat flux rises up to its maximum value (at $t = 0$ ms), while the wall normal velocity is positive and starts to decrease. At the same time, the flame moves towards the wall. It can been seen that the velocity is nearly zero when the highest heat flux appears, which is higher than the laminar SWQ case (black asterisk). After that, the velocity drops under zero, it stays negative, and the heat flux decreases again. The relation of the velocity and the temperature for the SWQ cases and the HOQ flame at characteristic points in time are plotted in Figure 22.

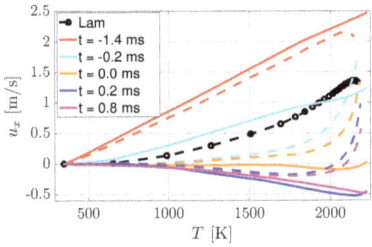

Figure 22. Temporal evolution of the flame. The solid line marks the HOQ and the dashed line the SWQ case.

This kind of representation decouples the dependency of the spacial coordinate. Before the flame interacts with the wall ($t = -1.4$ ms, solid red line), the flame has a linear dependency, as the temperature increases the velocities increases, too. Right before the flame quenches ($t = -0.2$ ms), the velocity over the whole flame is reduced. Like seen before in Figure 21, at the moment of quenching, the velocity is very small. The high heat fluxes appear because the velocities between the flame and the wall decreases if the flame reaches the wall (Figure 21b,c), due to the fact that the flame can closely approach the wall, which leads to high temperature gradients. As already mentioned and known from the literature, the laminar wall heat flux is lower and the quenching distance is greater, which can be seen in Figure 21a,c. Due to the higher wall normal velocity in the laminar case, the convection away from the wall is greater, which explains why this flame is farther away from the wall (Figure 21b). The $T - u_x$ plot (Figure 22) reveals that the velocity in the laminar case is much greater than zero over most of the flame. Finally, the behavior of the turbulent SWQ flame will be discussed. The temporal evolution of the heat flux, the wall normal velocity and the flame position are also shown in Figure 21. Qualitatively, the shapes are similar to the one-dimensional HOQ flame. Certainly, the increase of the wall heat flux and the decrease of the velocity start earlier in the turbulent case because this flame is closer to the wall. The behavior right before the quenching occurs ($t = -0.2$ ms) is different too because the turbulent SWQ flame moves much slower. Furthermore, the linear dependency of the velocity from the temperature vanishes much faster for the SWQ flame; in the moment before quenching, it is gone. As it can be seen, at the moment of quenching ($t = 0$ ms), the velocity of the HOQ is nearly zero over the whole temperature range, while, for the SWQ flame, the velocity is very small in the range of $T = 300$ K–$T = 1750$ K. After quenching, both flames have the same behavior for the region between the wall and the center of the flame. The drop of the velocity is also visible in the 3D plots in Figures 19 and 20. As observed in these figures, the fluid moves away from the tracked point and, as a result, the wall normal velocity gets lower than in the laminar case. During this process, the convex shaped flame front gets flattened. As a result, the turbulent flame behaves locally like an HOQ flame because the effects of the velocity are getting negligible. The maximum wall heat flux is $\Phi_{w,max} = 311$ kW/m^2 and the minimal wall distance is $x_{min} = 0.475$ mm, which are very close to the HOQ flame ($\Phi_{w,max,HOQ} = 313$ kW/m^2, $x_{min,HOQ} = 0.417$ mm). After the turbulent flame reached the maximum $\Phi_{w,max}$ ($t > 0$ ms), the profiles from Figure 21 are nearly the same again. However, a difference can be seen in the $T - u_x$ plot (Figure 22), for the region far away from the wall, the velocity of the turbulent flame is much greater because the fluid of the SWQ flame can accelerate again.

With the knowledge gained above, the identified regimes are very natural and somehow obvious, but the characterization raises the question if also a jump-like downstream movement could appear. Therefore, pockets of fresh gases must appear downstream of the flame front inside the burnt region. In our configuration, this seems to be very unlikely and was not observed, but it could be possible e.g., if large eddies transport fresh gases into the burnt region. Due to the flame topology, these events would not be connected with high heat fluxes because the fresh gases do not preheat fluid near the walls.

4. Conclusions

This work was conducted to gain insight into the flame–wall–turbulence interaction. The analysis utilized results obtained from the simulation of a well-suited SWQ configuration where the statistical data was confirmed to be in good agreement with measurements. Building upon this predicting capability, the availability of the highly resolved, three-dimensional transient field gave access to the physical processes like the heat flux evolution that significantly depart from the laminar operation. First, several processes were identified that show a distinct phenomenological behavior associated with the quenching point displacement. Specifically, the three archetypes being the downstream, upstream and jump-like movements were shown to operate on individual time-scales and heat flux evolutions. This physical understanding was then supplemented with statistical evaluations confirming its generality. For this, the conditionally averaged data showed that the average downstream movement of the quenching point yields significantly lower heat fluxes than the laminar SWQ flame and vice versa for the upstream movement. Significantly larger heat fluxes were observed when the quenching point jumped. The root causes for this were subsequently identified by considering the full interaction of the turbulent field, the flame and the wall. It revealed that the near-wall flame wrinkling by turbulent structures locally creates a departure from the SWQ towards the HOQ regime. This latter is known for conditions that enable the flame to approach the wall very rapidly causing a strong interaction with high heat fluxes. We showed that, by the curvature of the flame towards the wall, these conditions can even get surpassed by a flame-tangential divergence of the velocity field.

The identification of the mechanisms of FTWI was conducted based on simulations of a laboratory scale experimental SWQ burner. The configuration was chosen first, since its geometrical complexity is a reasonable compromise in between computational effort and practical relevance regarding the process of interest and, second, basic validation data are available. With respect to the generality of the results and their relevance for future combustor design, it should be mentioned that the main goal of this work was to show that simulations with the given approaches can handle such configurations and their results provide a sound basis for an understanding of the physical processes. Considering the phenomenological character of our analysis, it is certainly transferable for in terms of the fundamental FTWI mechanism. However, the parameter specific to a configuration will decide how they scale and accordingly quantitatively behave. In this regard, the Reynolds number is generally higher in real applications, intensifying the FTWI. Accordingly, the mechanism of regime alteration (SWQ-HOQ) will likewise be enhanced. Furthermore, for regimes of higher Karlovitz numbers, stretch might further modify the near wall flame structure, causing significantly different peaks of the heat release. Likewise, the specific fuel is of major importance e.g., with respect to the pollutants formed in the walls vicinity. Similarly, the wall treatment (e.g., effusion cooling) will have an influence. In this regard, we mentioned that the tabulation approach should be chosen with care depending on the quantities of interest.

Author Contributions: A.H. performed the simulations. A.H., G.K. and S.G. exploited the numerical data and analyzed the results. A.H. wrote the paper, while G.K. and C.H. further supported improving the manuscript. J.J. contributed by providing materials, computing resources and supervision.

Funding: The authors gratefully acknowledge the financial support by the Deutsche Forschungsgemeinschaft (DFG) through grants SFB/TRR 150 and in the framework of the Excellence Initiative, Darmstadt Graduate School of Energy Science and Engineering (GSC 1070).

Acknowledgments: All computations were performed on the Lichtenberg High Performance Computer (HHLR) at the TU Darmstadt.

Conflicts of Interest: The authors declare no conflict of interest.

References

1. International Energy Agency (IEA). *World Energy Outlook*; IEA: Paris, France, 2017.
2. BP. *Statistical Review of World Energy*; BP: London, UK, 2018.
3. Lazik, W.W.; Doerr, T.; Bake, S.S.; vd Bank, R.R.; Rackwitz, L.L. Development of Lean-Burn Low-NOx Combustion Technology at Rolls-Royce Deutschland. In *ASME Turbo Expo: Power for Land, Sea, and Air*; Volume 3: Combustion, Fuels and Emissions, Parts A and B; American Society of Mechanical Engineers (ASME): New York, NY, USA, 2008; pp. 797–807. [CrossRef]
4. Dreizler, A.; Böhm, B. Advanced laser diagnostics for an improved understanding of premixed flame–wall interactions. *Proc. Combust. Inst.* **2015**, *35*, 37–64. [CrossRef]
5. Poinsot, T.; Veynante, D. *Theoretical and Numerical Combustion*, 3rd ed.; Thierry Poinsot and Denis Veynante; Institut de Mécanique des Fluides de Toulouse: Toulouse, France, 2012.
6. Cheng, R.; Bill, R.; Robben, F. Experimental study of combustion in a turbulent boundary layer. *Symp. (Int.) Combust.* **1981**, *18*, 1021–1029. [CrossRef]
7. Gruber, A.; Sankaran, R.; Hawkes, E.R.; Chen, J.H. Turbulent flame–wall interaction: A direct numerical simulation study. *J. Fluid Mech.* **2010**, *658*, 5–32. [CrossRef]
8. Saffman, M. Parametric studies of a side wall quench layer. *Combust. Flame* **1984**, *55*, 141–159. [CrossRef]
9. Lu, J.; Ezekoye, O.; Greif, R.; Sawyer, R. Unsteady heat transfer during side wall quenching of a laminar flame. *Symp. (Int.) Combust.* **1991**, *23*, 441–446. [CrossRef]
10. Ezekoye, O.; Greif, R.; Sawyer, R. Increased surface temperature effects on wall heat transfer during unsteady flame quenching. *Symp. (Int.) Combust.* **1992**, *24*, 1465–1472. [CrossRef]

11. Boust, B.; Sotton, J.; Labuda, S.; Bellenoue, M. A thermal formulation for single-wall quenching of transient laminar flames. *Combust. Flame* **2007**, *149*, 286–294. [CrossRef]
12. Alshaalan, T.M.; Rutland, C.J. Turbulence, scalar transport, and reaction rates in flame–wall interaction. *Symp. (Int.) Combust.* **1998**, *27*, 793–799. [CrossRef]
13. Alshaalan, T.M.; Rutland, C.J. Wall heat flux in turbulent premixed reacting flow. *Combust. Sci. Technol.* **2002**, *174*, 135–165. [CrossRef]
14. Jainski, C.; Rißmann, M.; Böhm, B.; Dreizler, A. Experimental investigation of flame surface density and mean reaction rate during flame–wall interaction. *Proc. Combust. Inst.* **2016**, *36*, 1827–1834. [CrossRef]
15. Jainski, C.; Rißmann, M.; Böhm, B.; Janicka, J.; Dreizler, A. Sidewall quenching of atmospheric laminar premixed flames studied by laser-based diagnostics. *Combust. Flame* **2017**, *183*, 271–282. [CrossRef]
16. Heinrich, A.; Ganter, S.; Kuenne, G.; Jainski, C.; Dreizler, A.; Janicka, J. 3D Numerical Simulation of a Laminar Experimental SWQ Burner with Tabulated Chemistry. *Flow Turbul. Combust.* **2017**, *100*, 535–559. [CrossRef]
17. Heinrich, A.; Ries, F.; Kuenne, G.; Ganter, S.; Hasse, C.; Sadiki, A.; Janicka, J. Large Eddy Simulation with tabulated chemistry of an experimental sidewall quenching burner. *Int. J. Heat Fluid Flow* **2018**, *71*, 95–110. [CrossRef]
18. Lehnhäuser, T.; Schäfer, M. Improved linear interpolation practice for finite-volume schemes on complex grids. *Int. J. Numer. Methods Fluids* **2002**, *38*, 625–645. [CrossRef]
19. Smagorinsky, J. General Circulation Experiments with the Primitive Equations, 1. the basic experiment. *Mon. Weather Rev.* **1963**, *91*, 99–164. [CrossRef]
20. Germano, M.; Piomelli, U.; Moin, P.; Cabot, W.H. A dynamic subgrid-scale eddy viscosity model. *Phys. Fluids A Fluid Dyn.* **1991**, *3*, 1760–1765. [CrossRef]
21. Lilly, D.K. A proposed modification of the Germano subgrid-scale closure method. *Phys. Fluids A Fluid Dyn.* **1992**, *4*, 633–635. [CrossRef]
22. Zhou, G.; Davidson, L.; Olsson, E. Transonic inviscid/turbulent airfoil flow simulations using a pressure based method with high order schemes. In Proceedings of the Fourteenth International Conference on Numerical Methods in Fluid Dynamics, Bangalore, India, 11–15 July 1994; Deshpande, S.M., Desai, S.S., Narasimha, R., Eds.; Springer: Berlin/Heidelberg, Germany, 1995; pp. 372–378.
23. Van Oijen, J.A.; de Goey, L.P.H. Modelling of Premixed Laminar Flames using Flamelet-Generated Manifolds. *Combust. Sci. Technol.* **2000**, *161*, 113–137. [CrossRef]
24. Smith, G.P.; Golden, D.M.; Frenklach, M.; Moriarty, N.W.; Eiteneer, B.; Goldenberg, M.; Bowman, C.T.; Hanson, R.K.; Song, S.; Gardiner, W.C., Jr.; et al. GRI-Mech 3.0, 1999. Available online: http://combustion.berkeley.edu/gri-mech/ (accessed on 1 September 2018).
25. Chem1D. A One-Dimensional Laminar Flame Code, Developed at Eindhoven University of Technology. Available online: https://www.tue.nl/en/university/departments/mechanical-engineering/research/research-groups/multiphase-and-reactive-flows/our-expertise/chem1d/ (accessed on 1 September 2018).
26. Van Oijen, J.A.; Lammers, F.A.; de Goey, L.P.H. Modeling of complex premixed burner systems by using flamelet-generated manifolds. *Combust. Flame* **2001**, *127*, 2124–2134. [CrossRef]
27. Ketelheun, A.; Kuenne, G.; Janicka, J. Heat Transfer Modeling in the Context of Large Eddy Simulation of Premixed Combustion with Tabulated Chemistry. *Flow Turbul. Combust.* **2013**, *91*, 867–893. [CrossRef]
28. Kuenne, G.; Euler, M.; Ketelheun, A.; Avdić, A.; Dreizler, A.; Janicka, J. A Numerical Study of the Flame Stabilization Mechanism Being Determined by Chemical Reaction Rates Submitted to Heat Transfer Processes. *Zeitschrift für Physikalische Chemie* **2014**, *229*, 643–662. [CrossRef]
29. Avdić, A.; Kuenne, G.; Janicka, J. Flow Physics of a Bluff-Body Swirl Stabilized Flame and their Prediction by Means of a Joint Eulerian Stochastic Field and Tabulated Chemistry Approach. *Flow Turbul. Combust.* **2016**, *97*, 1185–1210. [CrossRef]
30. Avdić, A.; Kuenne, G.; di Mare, F.; Janicka, J. LES combustion modeling using the Eulerian stochastic field method coupled with tabulated chemistry. *Combust. Flame* **2017**, *175*, 201–219. [CrossRef]
31. Meier, T.; Kuenne, G.; Ketelheun, A.; Janicka, J. Numerische Abbildung von Verbrennungsprozessen mit Hilfe detaillierter und tabellierter Chemie. In *VDI Berichte*; Number 2162; VDI Verlag GmbH: Duisburg, Germany, 2013; pp. 643–652.
32. Ganter, S.; Heinrich, A.; Meier, T.; Kuenne, G.; Jainski, C.; Rißmann, M.C.; Dreizler, A.; Janicka, J. Numerical analysis of laminar methane–air side-wall-quenching. *Combust. Flame* **2017**, *186*, 299–310. [CrossRef]

33. Ganter, S.; Straßacker, C.; Kuenne, G.; Meier, T.; Heinrich, A.; Maas, U.; Janicka, J. Laminar near-wall combustion: Analysis of tabulated chemistry simulations by means of detailed kinetics. *Int. J. Heat Fluid Flow* **2018**, *70*, 259–270. [CrossRef]
34. Williams, F.A. *Combustion Theory: The Fundamental Theory of Chemically Reacting Flow Systems*; Addison/Wesley Pub. Co.: Boston, MA, USA, 1985; p. 680.
35. Colin, O.; Ducros, F.; Veynante, D.; Poinsot, T. A thickened flame model for large eddy simulations of turbulent premixed combustion. *Phys. Fluids* **2000**, *12*, 1843–1863. [CrossRef]
36. Butler, T.; O'Rourke, P. A numerical method for two dimensional unsteady reacting flows. *Symp. (Int.) Combust.* **1977**, *16*, 1503–1515. [CrossRef]
37. Durand, L.; Polifke, W. Implementation of the Thickened Flame Model for Large Eddy Simulation of Turbulent Premixed Combustion in a Commercial Solver. In *ASME Turbo Expo: Power for Land, Sea, and Air*; American Society of Mechanical Engineers (ASME): New York, NY, USA, 2007; Volume 2, pp. 869–878.
38. Kuenne, G.; Ketelheun, A.; Janicka, J. LES modeling of premixed combustion using a thickened flame approach coupled with FGM tabulated chemistry. *Combust. Flame* **2011**, *158*, 1750–1767. [CrossRef]
39. Charlette, F.; Meneveau, C.; Veynante, D. A power-law flame wrinkling model for LES of premixed turbulent combustion Part I: Non-dynamic formulation and initial tests. *Combust. Flame* **2002**, *131*, 159–180. [CrossRef]
40. Peters, N. Laminar flamelet concepts in turbulent combustion. *Symp. (Int.) Combust.* **1986**, *21*, 1231–1250. [CrossRef]
41. Van Oijen, J.; Bastiaans, R.; de Goey, L. Low-dimensional manifolds in direct numerical simulations of premixed turbulent flames. *Proc. Combust. Inst.* **2007**, *31*, 1377–1384. [CrossRef]
42. Van Oijen, J.; Groot, G.; Bastiaans, R.; de Goey, L. A flamelet analysis of the burning velocity of premixed turbulent expanding flames. *Proc. Combust. Inst.* **2005**, *30*, 657–664. [CrossRef]

© 2018 by the authors. Licensee MDPI, Basel, Switzerland. This article is an open access article distributed under the terms and conditions of the Creative Commons Attribution (CC BY) license (http://creativecommons.org/licenses/by/4.0/).

MDPI
St. Alban-Anlage 66
4052 Basel
Switzerland
Tel. +41 61 683 77 34
Fax +41 61 302 89 18
www.mdpi.com

Fluids Editorial Office
E-mail: fluids@mdpi.com
www.mdpi.com/journal/fluids

www.ingramcontent.com/pod-product-compliance
Lightning Source LLC
LaVergne TN
LVHW070603100526
838202LV00012B/549